SpringerBriefs in Applied Sciences and Technology

SpringerBriefs present concise summaries of cutting-edge research and practical applications across a wide spectrum of fields. Featuring compact volumes of 50 to 125 pages, the series covers a range of content from professional to academic.

Typical publications can be:

- A timely report of state-of-the art methods
- An introduction to or a manual for the application of mathematical or computer techniques
- A bridge between new research results, as published in journal articles
- A snapshot of a hot or emerging topic
- An in-depth case study
- A presentation of core concepts that students must understand in order to make independent contributions

SpringerBriefs are characterized by fast, global electronic dissemination, standard publishing contracts, standardized manuscript preparation and formatting guidelines, and expedited production schedules.

On the one hand, **SpringerBriefs in Applied Sciences and Technology** are devoted to the publication of fundamentals and applications within the different classical engineering disciplines as well as in interdisciplinary fields that recently emerged between these areas. On the other hand, as the boundary separating fundamental research and applied technology is more and more dissolving, this series is particularly open to trans-disciplinary topics between fundamental science and engineering.

Indexed by EI-Compendex, SCOPUS and Springerlink.

Abdallah Mohamed Hamed

Studies on the Confocal Laser Microscope

 Springer

Abdallah Mohamed Hamed ⓘ
Department of Physics, Faculty of Science
Ain Shams University
Cairo, Egypt

ISSN 2191-530X   ISSN 2191-5318   (electronic)
SpringerBriefs in Applied Sciences and Technology
ISBN 978-3-031-87274-7   ISBN 978-3-031-87275-4   (eBook)
https://doi.org/10.1007/978-3-031-87275-4

This Springer imprint is published by the registered company Springer Nature Switzerland AG
The registered company address is: Gewerbestrasse 11, 6330 Cham, Switzerland

If disposing of this product, please recycle the paper.

I dedicate this book to the spirit of my parents

Preface

The confocal microscope consists of two objectives arranged in tandem, and the object is found in the common short focus corresponding to the two objectives. The coherent illumination provided by the laser is rendered parallel to the first objective, which is limited by an aperture P_1, and the light transmitted from the second objective is limited by an aperture P_2 incident upon a point coherent detector. The image is built where the scanned object is synchronized with the electronic scanning in the detection plane. The theory showed that the image intensity in the detection plane is the modulus square of the convolution product of the object and the resultant point spread function (RPSF). The RPSF is computed from the product of the PSF corresponding to each objective lens. In addition, the coherent transfer function (CTF) is the convolution product of the two apertures corresponding to the two objectives.

In this book, we review some modulated apertures and compute the corresponding point spread functions (PSFs) to improve microscope resolution in Chaps. 1 and 2. These apertures have linear, quadratic, concentric black and white (B/W) zones. In addition, linear-quadratic and polynomial apertures are investigated. The other apertures have Hamming, Cauchy, rectangular, and hexagonal shapes. These apertures are applied to a confocal scanning laser microscope (CSLM). The computation of coherent transfer functions (CTFs) for some modulated apertures is presented in Chap. 3. The imaging of microscopic objects using a confocal microscope is investigated in Chap. 4.

In addition, a theoretical study on a coherent non-scanned laser microscope (CNSM) was conducted in Chap. 5. The lateral and axial point spread functions in confocal imaging systems using binary amplitude masks were computed in Chap. 6, and the misalignment errors combined with wavefront aberrations using linear and quadratic apertures were investigated in Chap. 7. We calculated the diffraction intensity using a confocal microscope with a laterally displaced truncated Gaussian aperture in Chap. 8. I extended the results of Marechal microscopy to confocal scanning microscopy. In Chaps. 9 and 10, we study the spatial coherence for the confocal optical systems using quadratic, and concentric B/W apertures. Finally, in Chap. 11, we present an application to process cardiac images using cardiac apertures in CSLM.

Abdallah Mohamed Hamed
Department of Physics
Faculty of Science
Ain Shams University
Cairo, Egypt

Competing Interests The author has no competing interests to declare that are relevant to the content of this manuscript.

Contents

Chapter 1
Basics of Confocal Scanning Laser Microscope

1.1 Introduction

In scanning optical confocal microscopy, the quality of both lenses, the objective lens and collector lens, is equally responsible for the point spread function. If the object moves along the optical axis, the energy decay versus the defect of focus follows a sinc^4 function, as given for the first time by NOMARSKI [1, 2].

It was shown early by Sheppard et al. [3–7] that the resolution can be improved by using an annular aperture compared with an open circular aperture. The microscope resolution is dependent on the wavelength of illumination and the numerical aperture NA or the aperture size for a certain focal length; hence, the theoretical limit of resolution is computed as follows: resolution $= \lambda/\text{NA}$. The distribution of the aperture has little effect on the resolution and contrast Hamed et al. [8–11]. The resolution and contrast enhancement in confocal microscopes are discussed in [12–16]. Scanning twice is suggested in [17] for better resolution. The coherent transfer function and its significance for image scanning microscopy are discussed.

In the following, the PSF corresponding to some modulated apertures are discussed in Chap. 2, for the sake of improving the resolution of the confocal microscope. In addition, the CTF for modulated apertures is computed in Chap. 3.

1.2 Images Were Acquired with a Confocal Laser Scanning Microscope

First, consider a coherent imaging system with an amplitude point spread function (PSF) represented as $h(x, y)$ and an object with amplitude transmittance $g(x, y)$. In this case, the image intensity is given by [4–8]:

© The Author(s), under exclusive license to Springer Nature Switzerland AG 2025
A. M. Hamed, *Studies on the Confocal Laser Microscope*,
SpringerBriefs in Applied Sciences and Technology,
https://doi.org/10.1007/978-3-031-87275-4_1

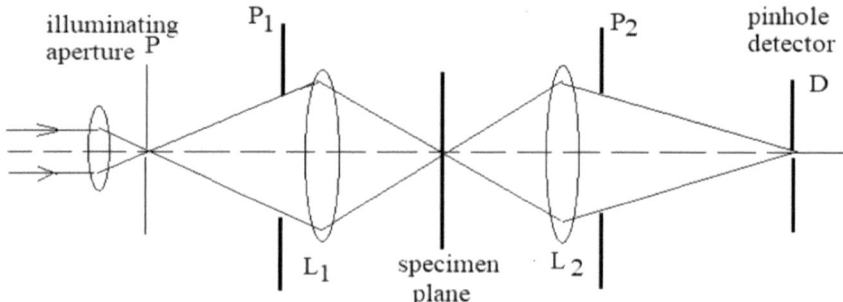

Fig. 1.1 Image formation under a transmission confocal microscope

$$I(x, y) = \left| \iint\limits_{-\infty}^{\infty} g(x', y')h(x - x', y - y')dx'dy' \right|^2 \qquad (1.1)$$

In a symbolic form, it is written as follows:

$$I(x, y) = |g(x, y) \otimes h(x, y)|^2 \qquad (1.2)$$

Where $\otimes$ represents the convolution operation product of the object and the PSF of the objective lens in the case of the conventional microscope.

In a confocal microscope working in transmission, as shown in Fig. 1.1, the intensity at a point (x_d, y_d) in the detector plane is given by the modulus square of the convolution $h_1(x, y)g(x, y) \otimes h_2(x, y)$. It is represented mathematically as follows:

$$I(x_d, y_d) = \left| \int \int h_1(x, y)g(x - x_s, y - y_s)h_2(x_d - x, y_d - y)dxdy \right|^2 \qquad (1.3)$$

where in Eq. (1.3), the scanned object is in the confocal plane of the objectives illuminated by the PSF of the first lens and the PSF of the second lens.

A pinhole is located at the center of the detector at $x_d = 0$ and $y_d = 0$; hence, Eq. (1.3) becomes:

$$I(x_d = 0, y_d = 0) = \left| \int \int h_1(x, y)h_2(-x, -y)g(x - x_s, -y_s)dx\,dy \right|^2 \qquad (1.4)$$

If h_2 is even, Eq. (1.4) is symbolically written as follows:

$$I(x, y) = |h_1(x, y)h_2(x, y) \otimes g(x, y)|^2 = |h_{\text{eff}}(x, y) \otimes g(x, y)|^2 \qquad (1.5)$$

Therefore, the confocal microscope behaves as a coherent imaging system with an effective amplitude point spread function h_{eff}, named the RPSF in CSLM. The

RPSF or h_{eff} given by the product of those for the illuminating and collecting lenses is as follows:

$$h_{\text{eff}}(x, y) = h_1(x, y) \times h_2(x, y) \tag{1.6}$$

1.2.1 The Resultant Point Spread Function (RPSF) in the CSLM

The point spread function (PSF) is the smear we captured by our microscope when the object is smaller than the resolution of our microscope is computed by operating F.T. upon the aperture located in the plane of Cartesian coordinates (u, v). The PSF is written in integral form as:

$$h(x, y) = \int\int_{-\infty}^{\infty} P(u, v) e^{\frac{-j2\pi}{\lambda f}(xu+yv)} du dv \tag{1.7}$$

$P(u, v)$, λ and f are the Pupil function, wavelength, and focal length respectively. Equation (1.7) is written in symbolic form as follows:

$$h(x, y) = \text{F.T.}\{P(u, v)\} \tag{1.8}$$

Pupil function is a function that describes how a light wave is affected upon transmission through an optical imaging system such as a camera, microscope, or human eye, and it is an important tool for studying optical imaging systems and their performance.

Since the Resultant Point Spread Function (RPSF) in the (CLSM) or h_{eff} is computed from the simple product of $h_1 \cdot h_2$, Eq. (1.6), the image of a point object is deduced by taking the modulus square of Eq. (1.6).

$$I(x, y) = |h_1(x, y) \cdot h_2(x, y)|^2 = |h_{\text{eff}}(x, y)|^2 \tag{1.9}$$

Equation (1.9) is deduced from Eq. (1.5) by substituting the point object

$$g(x, y) = \delta(x, y)$$

In the case of a point object, this can be represented by the Dirac delta distribution, i.e., $g(x, y) = \delta(x, y)$, and formula (1.5) gives:

$$I(x, y) = |h_{\text{eff}}(x, y) \otimes \delta(x, y)|^2 = |h_{\text{eff}}(x, y)|^2 = |h_1(x, y) \times h_2(x, y)|^2 \tag{1.10}$$

Each of these points spread functions can be calculated by the knowledge of their corresponding pupil distributions as follows:

$$h_1(x, y) = F.T.[P_1(u, v)], h_2(x, y) = F.T.[P_2(u, v)] \tag{1.11}$$

where (u, v) are the Cartesian coordinates in the aperture plane, and F.T. denotes the Fourier transform operation.

If the objective lens is circular, the PSF or the amplitude impulse response can be easily calculated as follows [14]:

$$h_1(r) = const.J_1[(2\pi ar/(\lambda f)/(2\pi ar/(\lambda f)] \tag{1.12}$$

where $r^2 = x^2 + y^2$, , and r is the radial coordinate in the Fourier plane (x, y), as shown in Fig. 1.1.

Equation (1.12) can be rewritten as follows:

$$h_1(z) = const.[J_1(z)/(z)] \tag{1.13}$$

where $z = 2\pi\rho_0 r/\lambda f$ is the reduced coordinate, $NA = \rho_0/f$ is the numerical aperture of the objective lens, and $\rho_0 = a =$ is the maximum value of the aperture radius.

The RPSF can be easily calculated as follows:

$$h_r = h_1(z).h_2(z) = const.\left[\frac{J_1(z)}{z}\right]^2 \tag{1.14}$$

The corresponding image of a point is the $\left[\frac{J_1(z)}{z}\right]$ to the fourth power compared to the squared value of $\left[\frac{J_1(z)}{z}\right]$ In the case of a conventional microscope.

$$\mathbf{I}_{conv.}(\mathbf{z}) = \mathbf{const.}\left|\frac{\boldsymbol{J}_1(z)}{\boldsymbol{z}}\right|^2$$

while

$$\mathbf{I}_{conf}(\mathbf{z}) = \mathbf{const.}\left|\frac{\boldsymbol{J}_1(z)}{\boldsymbol{z}}\right|^4$$

For two symmetric objectives of circular apertures.

1.3 Coherent Transfer Function (CTF) Determined by Confocal Microscopy

It is known that the CTF is computed for the CSLM from the convolution product of the objective lenses of the confocal arrangement [9–11] since both objectives contribute equally to the resolution of the microscope. This is attributed to the manipulation of the point source and point detector in the microscope. Hence, the CTF is the function $c(u, v)$ computed as follows:

$$c(u, v) = P_1(u, v) \otimes P_2(u, v) \tag{1.15}$$

The CTF is computed directly or by using the Fourier transform (F.T.) technique [4].

The pupil function can be written as $P(u, v) = T(u, v)e^{\frac{i2\pi W}{\lambda}}$, where T is the transmission over the pupil, e is the natural logarithm base, W is the aberration function, which expresses the wavefront deviation in wavelengths, and (u, v) is the coordinates in the pupil plane. The factor $e^{\frac{i2\pi W}{\lambda}}$ Defines phase variation over the pupil in complex exponential notation. These two parts of the pupil function are not directly compatible, with the transmission T directly expressing field (amplitude) in the pupil, and the phase factor. $e^{\frac{i2\pi W}{\lambda}}$ expressing the phase. Given in units of wavelength, W numerically corresponds to the phase angle φ in units of the 2π (radians) full phase. Exenteration errors combined with wavefront aberrations [11, 18–21] are discussed in Chap. 6.

References

1. M. Minsky, U.S. Patent 3013467, Microscopy apparatus Dec. 19 (1961)
2. G. Nomarski, J. Opt. Soc. Am. **65**, 1166 (1975)
3. C.J.R. Sheppard, A. Choudhary, Image formation in the scanning microscope. Opt. Acta **24**, 1051–1073 (1977)
4. I.J. Cox, C.J.R. Sheppard, T. Wilson, Improvement in resolution by confocal microscopy. Appl. Opt. **21**, 778–781 (1982)
5. C.J.R. Sheppard, X.Q. Mao, Confocal microscopes with slit apertures. J. Mod. Opt. **35**, 1169–1185 (1988)
6. C.J.R. Sheppard, Super-resolution in confocal imaging. Optik **80**, 53–54 (1988)
7. G. Cox, C.J.R. Sheppard, Practical limits of resolution in confocal and nonlinear microscopy. Microsc. Res. Tech. **63**, 18–22 (2004)
8. J.J. Clair, A.M. Hamed, Theoretical studies on optical coherent microscope. Optik **64**, 133–141 (1983)
9. A.M. Hamed, J.J. Clair, Image and super-resolution in optical coherent microscopes. Optik **64**, 272–284 (1983)
10. A.M. Hamed, J.J. Clair, Studies on optical properties of confocal scanning optical microscope using pupils with radially transmission distribution. Optik **65**, 209–218 (1983)
11. A.M. Hamed, Aberration studies utilizing an optoelectronic coherent microscope. Optik **67**, 279–290 (1984)
12. N. Chen, C.H. Wong, C.J.R. Sheppard, Focal modulation microscopy. Opt. Express **16**(23), 18764–18769 (2008)

13. W. Gong, Ke. Se, N. Chen, C.J.R. Sheppard, Improved spatial resolution in fluorescence focal modulation microscopy. Optics Lett. **34**(22), 3508–3510 (2009). https://doi.org/10.1364/OL.34.003508

14. Y. Fang et al., Resolution and contrast enhancements of optical microscope based on point spread function engineering. Frontiers Optoelectronics **8**, 152–162 (2015)

15. G. Boyer, V. Sarafis, Two pinhole super-resolution using spatial filters. Optik **112**, 177–179 (2001)

16. Y. Fang et al., Enhancing the resolution and contrast in CW-STED microscopy. Optics Commun. **322**, 169–174 (2014)

17. G.M.R. De Lucia et al., Re-scan confocal microscopy: scanning twice for better resolution. Biomed. Opt. Express **4**, 2644–2656 (2013)

18. C.J.R. Sheppard, M. Gu, Improvement of axial resolution in confocal microscopy using annular pupil. Opt. Communication. **84**, 7–13 (1991)

19. M. Gu et al., Optimization of axial resolution in confocal imaging using annular pupils. Optik **93**, 87–90 (1993)

20. A.M. Hamed, Exenteration errors combined with wavefront aberration in a coherent scanning microscope. Optik **82**, 1–4 (1989)

21. A.M. Hamed, Computation of the lateral and axial point spread functions in confocal imaging systems using binary amplitude mask. J. Phys. (Pramana) **66**, 1037–1048 (2006)

Chapter 2
The Point Spread Function (PSF) for Some Modulated Apertures

In this chapter, we compute the point spread function corresponding to different aperture apodizations. We start with uniform circular and annular apertures followed by black and white concentric apertures. Then, we compute the PSF corresponding to the conic, linear, quadratic, and higher-order apertures. In addition, we computed the PSF corresponding to the Hamming, Cauchy, polynomial, and elliptic apertures. Finally, we computed the PSF corresponding to the rectangular, and hexagonal apertures.

2.1 Circular Aperture

The circular aperture of the transmission function is defined as follows:

$$P(u, v) = 1 \ \text{ for } \rho < \rho_0$$
$$= 0 \text{ for } \rho \geq \rho_0 \tag{2.1}$$

The PSF is written in integral form as:

$$h(x, y) = \int\limits_{-\infty}^{\infty}\!\!\int P(u, v) e^{\frac{-j2\pi}{\lambda f}(xu+yv)} du dv \tag{2.2}$$

Applying the two-dimensional Bessel-Fourier transform leads us to write the point spread function as follows:

$$h(r) = 2 \int\limits_{0}^{1}\int\limits_{0}^{2\pi} \exp\left(-j2\pi \frac{\rho\, r}{\lambda f} \cos\theta\right) \rho\, d\rho d\theta$$

© The Author(s), under exclusive license to Springer Nature Switzerland AG 2025
A. M. Hamed, *Studies on the Confocal Laser Microscope*,
SpringerBriefs in Applied Sciences and Technology,
https://doi.org/10.1007/978-3-031-87275-4_2

$$= 2\pi \int_0^1 J_0\left(2\pi \frac{\rho\, r}{\lambda f}\right) \rho\, d\rho \tag{2.3}$$

with $k = 2\pi/\lambda$, which is the propagation constant and $\rho_0 = 1$.

We obtained the impulse response for the objective lens or the PSF as follows:

$$h(w) = 2\left(\frac{f}{kr}\right)^2 \int_0^W w J_0(w)\, dw \tag{2.4}$$

with $W = k\, r/f$, and $w = k\, \rho\, r/f$.

With the help of recurrence relations and using integration by partition [1, 2], we obtain:

$$h(w) = \left[\frac{2J_1(W)}{W}\right] \tag{2.5}$$

2.2   Black and White Concentric Aperture

The pupil under investigation is an obstructed circular pupil with black and transparent equal areas.

Referring to Fig. 2.1. We have a circular pupil having a numerical aperture (N.A. = $n' \sin\theta$), where θ is the half-angular aperture of the lens under consideration and n' is the refractive index of the object medium. In this case, the point spread function for a pupil of this form can be easily calculated analytically, if we take into consideration that each zone is computed from the difference between two consecutive circular pupils as follows:

$$h_l(w) = \text{F.T.}\{P(B/W)\} \tag{2.6}$$

B/W means here the succession of black and white (transparent) areas.

F.T. is the Fourier transform operation.

Equation (2.6) can be rewritten as follows:

$$h_1(w) = \text{F.T.}\left\{\sum_{n=1}^N \Delta P_{\rho n}\right\} = \text{F.T.}\{\Delta P_{\rho 1} + \Delta P_{\rho 2} + \cdots + \Delta P_{\rho n}\} \tag{2.7}$$

where, e.g., $\Delta P_{\rho 1} = P_{\rho 1'} - P_{\rho 1}$ as in Fig. 2.1, and N is the total number of transparent zones constituting the whole pupil.

Hence, operating the Fourier transformation over (2.7) gives for the point spread function this expression [3, 4]:

Fig. 2.1 Circular pupil
obstructed with a finite
number of successive black
and transparent areas

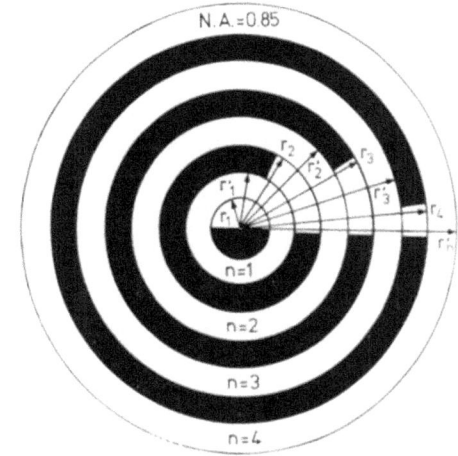

Fig. 2.2 Image intensity
distribution for a point object
using CSOM provided with
two identical pupils, type ρ^n,
for the objective lens and
collector lens

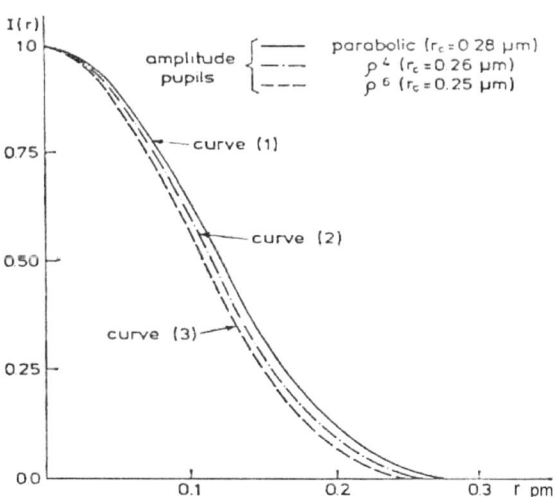

$$h(w) = \sum_{n=1}^{N} \left\{ r_n'^2 \left[\frac{2J_1(w_n')}{w_n'} \right] - r_n^2 \left[\frac{2J_1(w_n)}{w_n} \right] \right\} \tag{2.8}$$

with $w = 2\frac{r}{f}$, λ the wavelength of the illumination, f is the focal length of the lens,
and r is the radial coordinate in the object plane. This expression (2.8) is valid for
whatever the number of annuli constituting the pupil.

2.2.1 Special Case (Annular Aperture)

For annular aperture, using Eq. (2.8) for $n = 1$, we obtain the PSF as follows:

$$h_1(w) = \left[\frac{2J_1(w_n')}{w_n'}\right] - \varepsilon^2\left[\frac{2J_1(w_n)}{w_n}\right]\Bigg\} \tag{2.9}$$

where $\varepsilon = \frac{\rho_{int.}}{\rho_{ext.}}$; and $\rho_{int.}$, is the internal radius of the annulus, $\rho_{ext.} = \rho_0$, is the maximum radius of the circular aperture.

In addition, applying the two-dimensional Bessel-Fourier transform upon the annular aperture, as follows:

$$h_{annul.}(r) = 2\pi \int_0^\infty \delta(\rho - \rho_0)J_0\left(2\pi\rho\frac{r}{\lambda f}\right)\rho \, d\rho \tag{2.10}$$

We obtain the PSF as follows:

$$h_{annul.}(r) = 2J_0\left(2\pi\rho_0\frac{r}{\lambda f}\right) \tag{2.11}$$

2.3 Linear Aperture

We solve the problem in two dimensions, using polar coordinates, as follows:

The linear amplitude distribution for the pupil under consideration can be represented as follows:

$$\mathbf{P}(\rho) = \left|\frac{\rho}{\rho_0}\right|, \quad \text{for} \left|\frac{\rho}{\rho_0}\right| \leq 1 \tag{2.12}$$

This pupil is considered a hyper-resolving pupil because of its transmission distribution. It has the advantage of attenuating low frequencies; consequently, the diffracting object structure is imaged with enhanced contrast because the object areas of slowly varying transmission are attenuated [3, 4]. In this way, we believe that the low spatial frequency of the image is attenuated, which also provides better contrast of higher frequency components of the image spectrum.

Returning to formula (2.12), and applying the two-dimensional Bessel-Fourier transform, leads us to write the point spread function as follows:

$$h(\mathrm{r}) = 2 \int_0^1 \int_0^{2\pi} \rho \exp\left(-j2\pi \frac{\rho\, r}{\lambda f} \cos\theta \right) \rho\, d\rho\, d\theta$$

$$= 2\pi \int_0^1 \rho^2 J_0\left(2\pi \frac{\rho\, r}{\lambda f}\right) d\rho \tag{2.13}$$

with $k = 2\pi/\lambda$, which is the propagation constant and $\rho_0 = 1$.

We obtained the impulse response for the objective lens or the PSF as follows:

$$h_1(w) = 4\pi \left(\frac{f}{kr}\right)^3 \int_0^W w^2 J_0(w) dw \tag{2.14}$$

With $W = k\, r/f$, and $w = k\, \rho\, r/f$.

The solution of the integral (2.14) is given by Hamed [3, 4], during the treatment of conic amplitude distribution as a pupil function. This result leads to

$$h_1(w) = 4\left(\frac{f}{kr}\right)^3 \left[W^2 J_1(W) + W J_0(W) - 2 \sum_i J_i(W) \right] \tag{2.15}$$

(With $i = 1, 3, 5, \ldots$).

Hence, the impulse response of one pupil can be rewritten as follows:

$$h_1(w) = 4\left[\frac{J_1(W)}{W} + \frac{J_0(W)}{W^2} - 2 \sum_i J_i(W)/W^3 \right] \tag{2.16}$$

Consequently, the resulting impulse response of the optical system in a confocal microscope gives:

$$h_r(w) = h_1(W) h_2(W) = 16\pi^2 \left\{ \left[\frac{J_1(W)}{W} \right]^2 + \left[\frac{J_0(W)}{W^2} \right]^2 + 2 J_0(W) \left[\frac{J_1(W)}{W^3} \right] \right.$$

$$+ \left[2 \sum_i \frac{J_i(W)}{W^3} \right]^2$$

$$\left. -4 \left\{ \left[\frac{J_1(W)}{W} \right] + \left[\frac{J_0(W)}{W^2} \right] \right\} \cdot \sum_i \frac{J_i(W)}{W^3} \right\} \tag{2.17}$$

Hence, the intensity spread function gives the following:

$$I(W) = |h_1(W) \cdot h_2(W)|^2 = |h_r(W)|^2 \tag{2.18}$$

Formula (2.17) assumes two symmetric pupils: hence:

$$I(W) = 256\pi^4 \left[\frac{J_1(W)}{W} + \frac{J_0(W)}{W^2} - 2\sum_i J_i(W)/W^3 \right]^4 \qquad (2.19)$$

In the case of two points separated by a distance W_1, the image can be calculated easily to give:

$$I(W) = |[\delta_1(W) + \delta_2(W \pm W_1)] \otimes h_r(W)|^2$$
$$= |[h_r(W) + h_r(W \pm W_1)]|^2 \qquad (2.20)$$

In the case of the circular pupil for the objective lens, and a linear amplitude distribution for the collector lens, or vice versa, we obtain the total impulse response of the optical system as follows:

$$h_r(W) = 8\left[\frac{J_1(W)}{W}\right]\left[\frac{J_1(W)}{W} + \frac{J_0(W)}{W^2} - 2\sum_i \frac{J_i(W)}{W^3}\right] \qquad (2.21)$$

In the case of the annular pupil with the former pupil, which exhibits a radially increasing linear amplitude distribution, we obtain:

$$h_r(W) = \text{const.}\, J_0(W)\left[\frac{J_1(W)}{W} + \frac{J_0(W)}{W^2} - 2\sum_i \frac{J_i(W)}{W^3}\right] \qquad (2.22)$$

Consequently, the image of a point object, in this case, gives:

$$I(W) = \text{const.}\, J_0^2(W)\left[\frac{J_1(W)}{W} + \frac{J_0(W)}{W^2} - 2\sum_i \frac{J_i(W)}{W^3}\right]^2 \qquad (2.23)$$

2.4   Quadratic and Higher-Order Apertures

Suppose that we have a pupil with radial amplitude transmission, which varies according to the following formula:

$$P(\rho, \theta) = \rho^n, \ \text{with} |\rho| \leq \rho_0, \theta = \ \text{const.}, n = 2m, m = 0, 1, 2, 3, \ldots \qquad (2.24)$$

To improve the shape of the point spread function, and hence improve the resolution of the microscope, we expect to obtain a sharp peak for the PSF using increasing values of n.

The PSF, in the case of the pupil, is calculated with the help of Fourier techniques.

$$h_n(r) = \text{F.T.}\{P(\rho, \theta)\} \tag{2.25}$$

where F.T. represents the Fourier transform operation. In polar coordinates, the Fourier transformation given in formula (2.25) is written analytically as follows:

$$h_n(r) = \int_0^{2\pi} \int_0^1 \rho^n \exp\left[-jk\rho r \cos(\theta)\right] \rho \, d\rho \, d\theta \tag{2.26}$$

where we have considered that $\rho_0 = f = 1$.
Formula (2.26) is solved for θ to obtain:

$$h_n(r) = 2\pi \int_0^1 \rho^{n+1} J_0(k\rho r) d\rho \tag{2.27}$$

where k is the wavenumber and λ is the wavelength of light. With the change in variables, formula (2.27) gives:

$$h_n(r) = 2\left(\frac{1}{kr}\right)^{n+2} \int_0^w w^{n+1} J_0(w) dw \tag{2.28}$$

($w = k \rho r$, where $W = k r$ is the upper limit of the integration).
With the help of recurrence relations and using integration by partition [2, 5], we obtain:

$$h_n(W) = \text{constant}\left\{\frac{J_1(W)}{W} - n\frac{J_2(W)}{W^2} + n(n-2)\frac{J_3(W)}{W^3} - \ldots\right\} \tag{2.29}$$

Equation (2.29) is rewritten in series form as follows:

$$h_n(W) = \alpha_1 \text{ somb}(W) + \alpha_2 \sum_{i=2}^n \left\{(-1)^{i+1} \prod_{l=0}^{i-2}(n-2l)\frac{J_i(W)}{W^i}\right\} \tag{2.30}$$

α_1 and α_2 are constants, $i = 2, 3, 4\ldots, n$, and $l = 0, 1, 2, \ldots, n-2$
Formula (2.30) is the general analytical solution to obtain the coherent point spread function for the amplitude pupil.

(i) For $m = 0$, i.e., $n = 0$, we obtain the well-known solution for the PSF corresponding to a uniform circular aperture as follows:

$$h_n(W) = \text{constant}\frac{J_1(W)}{W} = \text{somb}(W); \text{ ref.}[2], \alpha = \text{const.} \tag{2.31}$$

Somb indicates the sombrero function.

(ii) For $m = 1$, i.e., $n = 2$, the PSF gives

$$h_2(w) = \text{const.} \left[\frac{J_1(w)}{w} - 2\frac{J_2(w)}{w^2} \right] \qquad (2.32)$$

Formula (2.32) gives the point spread function for a pupil having radially quadratic transmission (type ϱ^2).

(iii) Similarly, for $m = 2$,

$$h_4(w) = \text{const.} \left[\frac{J_1(w)}{w} - 4\frac{J_2(w)}{w^2} + 8\frac{J_3(w)}{w^3} \right] \qquad (2.33)$$

(iv) For $m = 3$, the point spread function is calculated to give

$$h_6(w) = \text{const.} \left[\frac{J_1(w)}{w} - 6\frac{J_2(w)}{w^2} + 24\frac{J_3(w)}{w^3} - 48\frac{J_4(w)}{w^4} \right] \qquad (2.34)$$

Computer programs calculate the PSF, and the image of a point object using formula (2.31) up to formula (2.34). The wavelength of light illumination was 6328 Å.

Figure 2.2 shows an image of a point object versus the radial coordinate r. Curve (1) corresponds to the case of two symmetric pupils of quadratic (parabolic) amplitude transmission, giving the first minimum diffraction pattern at $r_c = 0.28$ µm, which resembles the results of two annular pupils, while curve (2) gives r_c at 0.26 µm, which is given for two symmetric pupils of type ρ^4, and the last curve (3) gives r_c at 0.25 µm (pupils having ρ^6 amplitude transmission).

Figure 2.3 shows an image of a point object where curve (1), corresponds to a parabolic-annular combination giving r_c of 0.24 µm, while the second curve (2) drawn for the ρ^6 annular $\delta(\rho)$ combination gives an r_c of 0.225 µm. The last curve (3) was made for comparison with our results, giving r_c of 0.36 µm (uniform circular pupils). It is useful to mention here that the attenuation of the low spatial frequency of the pupil apertures of the imaging system improves the resolution of the CSOM (in the case of ρ^n or annular $\delta(\rho)$ pupils).

2.5 Linear-Quadratic Aperture

The aperture is composed of a quadratic distribution followed by a linear distribution and ends with a transparent uniform annulus, as shown in Fig. 2.4. This aperture is described as the sum of the following segments:

$$P_1(\rho) = \rho^2; \quad 0 \le |\rho| < |\rho_0/2| \text{ for the quadratic segment} \qquad (2.35)$$

Fig. 2.3 Image intensity distribution for a point object using CSOM provided with e^n pupil for the objective and annular pupil for the collector or vice versa

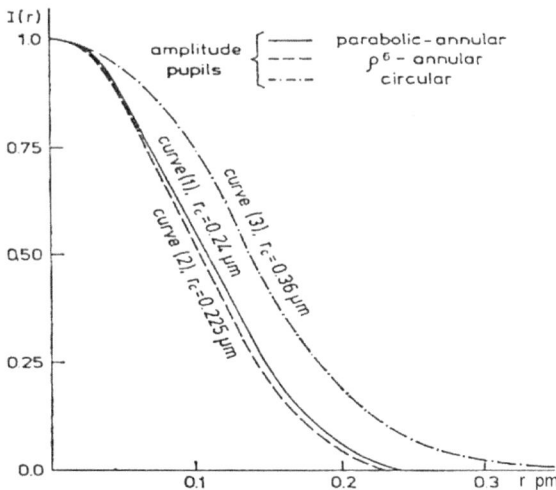

$$P_2(\rho) = \rho; \quad |\rho_0/2| \leq |\rho| < |3\rho_0/4| \text{ for the linear segment} \tag{2.36}$$

$$P_3(\rho) = 1; \quad |3\rho_0/4| \leq |\rho| < |\rho_0| \text{ for the annulus} \tag{2.37}$$

$$P_T = P_1(\rho) + P_2(\rho) + P_3(\rho) \tag{2.38}$$

This aperture is considered a hyper-resolving aperture since it attenuates the low spatial frequency such as the annular aperture, but it will gain better contrast than the annular, linear, or other modulated apertures by replacing the point detector with a quadrant detector [6]. This aperture was fabricated using the MATLAB code.

By applying the Fourier transform to the pupil function described in Eqs. (2.35–2.38), we obtain the PSF as follows (Fig. 2.4):

$$h(r) = 2 \int_0^{2\pi} \int_0^{\rho_0} P_T(\rho) \exp\left[\left(-\frac{j2\pi}{\lambda f}\right)\rho r \cos(\theta)\right] \rho \, d\rho \, d\theta \tag{2.39}$$

$$h(r) = 2 \int_0^{2\pi} \int_0^{\rho_0/2} \rho^2 \exp\left[\left(-\frac{j2\pi}{\lambda f}\right)\rho r \cos(\theta)\right] \rho \, d\rho \, d\theta$$

$$+ 2 \int_0^{2\pi} \int_{\rho_0/2}^{3\rho_0/4} \rho \exp\left[\left(-\frac{j2\pi}{\lambda f}\right)\rho r \cos(\theta)\right] \rho \, d\rho \, d\theta$$

$$+ 2 \int_0^{2\pi} \int_{3\rho_0/4}^{\rho_0} \exp\left[\left(-\frac{j2\pi}{\lambda f}\right)\rho r \cos(\theta)\right]\rho d\rho d\theta \qquad (2.40)$$

It is convenient to run the two-dimensional Fourier–Bessel transformation in polar coordinates to obtain:

$$h(r) = 4\pi \int_0^{\rho_0/2} \rho^2 J_0\left(\frac{2\pi\rho r}{\lambda f}\right)\rho d\rho + 4\pi \int_{\rho_0/2}^{3\rho_0/4} \rho J_0\left(\frac{2\pi\rho r}{\lambda f}\right)\rho d\rho$$

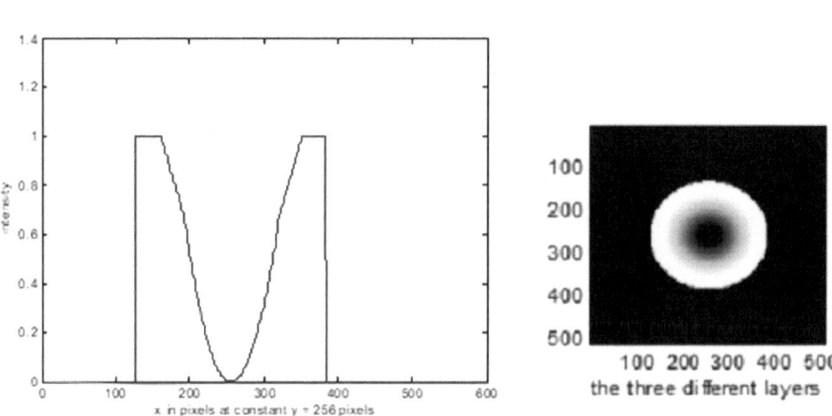

Fig. 2.4 The image (R.H.S.), and the corresponding line plot (L.H.S.) of the described aperture of three different segments. From the center, a quadratic distribution is followed by a linear distribution and ends with a transparent uniform annulus. The plot was taken at the horizontal central section at $y = 256$ pixels

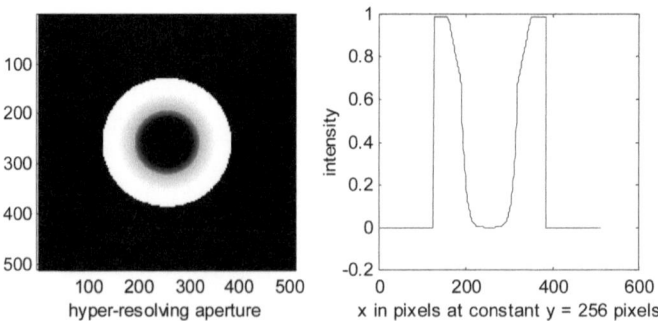

Fig. 2.5 On the left, the described aperture is shown with a total radius of 128 pixels, while on the right is a line plot of the aperture at a horizontal line at 256 pixels. The aperture has a central dark disk, ρ^6, linear, constant transmission segments arranged from the center with a ratio 0.1:0.4:0.25:0.25 of the radius

$$+ 4\pi \int_{3\rho_0/4}^{\rho_0} (1) J_0\left(\frac{2\pi\rho r}{\lambda f}\right) \rho \, d\rho \tag{2.41}$$

where J_0 is the zero-order Bessel function and $r = \sqrt{(x^2 + y^2)}$ Is the radial coordinate in the Fourier plane. We obtained this result for the PSF as follows [7]:

$$h_{1\text{st model}}(r) = 2\pi \left[\frac{J_1(w_0)}{w_0} - \frac{J_1(3w_0/4)}{3w_0/4}\right] + \frac{27\pi}{16} \left\{\left[\frac{J_1(w_2)}{w_2} - \frac{8J_1(w_1)}{27w_1}\right]\right.$$
$$+ \left[\frac{J_0(w_2)}{w_2^2} - \frac{8J_0(w_1)}{27w_1^2}\right] - 2\sum_i \left[J_i(w_2)/w_2^3 - \left(\frac{8}{27}\right)J_i(w_1)/w_1^3\right]\right\}$$
$$+ 2\pi \left[\frac{J_1\left(\frac{w_0}{2}\right)}{\frac{w_0}{2}} - \frac{2J_2\left(\frac{w_0}{2}\right)}{\left(\frac{w_0}{2}\right)^2}\right] \tag{2.42}$$

The 2nd model of the hyper-resolving aperture differs from the 1st model in that the quadratic segment is replaced by ρ^6 distributions of width 0.4 × radius and the central disk of radius = 0.1 × radius added, as shown in Fig. (2.5). The total radius = 128 pixels (Fig. 2.5).

$$P_1(\rho) = 0.0001; 0 \leq |\rho| < \left|\frac{\rho_0}{10}\right| \text{ for the central dark disk} \tag{2.43}$$

$$P_2(\rho) = \rho^6; |\rho_0/10| \leq |\rho| < |\rho_0/2| \text{ for the segment of } \rho^6 \tag{2.44}$$

$$P_3(\rho) = \rho; |\rho_0/2| \leq |\rho| < |3\rho_0/4| \text{ for the linear segment} \tag{2.45}$$

$$P_4(\rho) = 1; 4|3\rho_0/4| \leq |\rho| < |\rho_0| \text{ for the annulus} \tag{2.46}$$

$$P_T = P_1(\rho) + P_2(\rho) + P_3(\rho) + P_4(\rho) \tag{2.47}$$

Using Eqs. (2.43–2.47) and substituting Eq. (2.39) to obtain the PSF analogously to that of the 1st model, we obtain:

$$h(r) = 2 \int_0^{2\pi} \int_{\rho_0/10}^{\rho_0/2} \rho^6 \exp\left[\left(-\frac{j2\pi}{\lambda f}\right)\rho r \cos(\theta)\right] \rho \, d\rho \, d\theta$$
$$+ 2 \int_0^{2\pi} \int_{\rho_0/2}^{3\rho_0/4} \rho \exp\left[\left(-\frac{j2\pi}{\lambda f}\right)\rho r \cos(\theta)\right] \rho \, d\rho \, d\theta$$

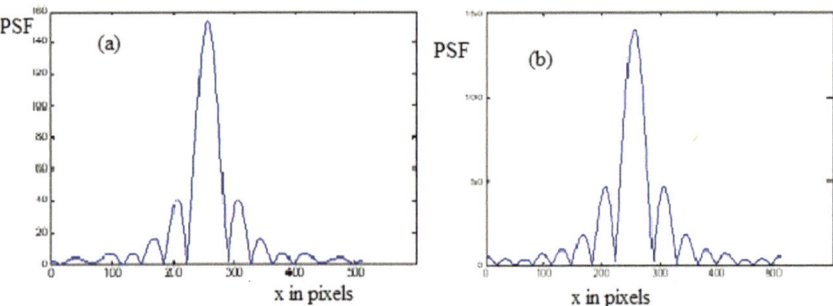

Fig. 2.6 The absolute value of the PSF corresponding to the two models of hyper-resolving apertures at the horizontal line at 256 pixels. (a) The PSF corresponding to the aperture of quadratic - linear - constant segments arranged from the center with a ratio of 2:1:1, while (b) the PSF corresponding to the 2nd aperture of dark - rho 6- linear-constant segments arranged from the center with a ratio of 0.1:0.4:0.25:0.25 for the radii

$$+ 2 \int_{0}^{2\pi} \int_{3\rho_0/4}^{\rho_0} \exp\left[\left(-\frac{j2\pi}{\lambda f}\right)\rho r \cos(\theta)\right] \rho \, d\rho \, d\theta \qquad (2.48)$$

Equation (2.48) was solved to obtain this result for the PSF [7]:

$$h_{\text{2nd model}}(r) = 2\pi \left[\frac{J_1(w_0)}{w_0} - \frac{J_1(3w_0/4)}{3w_0/4}\right] + \frac{27\pi}{16}\left\{\left[\frac{J_1(w_2)}{w_2} - \frac{8J_1(w_1)}{27w_1}\right]\right.$$

$$+ \left[\frac{J_0(w_2)}{w_2^2} - \frac{8J_0(w_1)}{27w_1^2}\right] - 2\sum_i\left[J_i(w_2)/w_2^3 - \left(\frac{8}{27}\right)J_i(w_1)/w_1^3\right]\right\}$$

$$+ 2\pi\left\{\left[\frac{J_1\left(\frac{w_0}{2}\right)}{\frac{w_0}{2}} - \frac{J_1\left(\frac{w_0}{10}\right)}{\frac{w_0}{10}}\right] - 6\left[\frac{J_2\left(\frac{w_0}{2}\right)}{\left(\frac{w_0}{2}\right)^2} - \frac{J_2\left(\frac{w_0}{10}\right)}{\left(\frac{w_0}{10}\right)^2}\right]\right.$$

$$+ 24\left[\frac{J_3\left(\frac{w_0}{2}\right)}{\left(\frac{w_0}{2}\right)^3} - \frac{J_3\left(\frac{w_0}{10}\right)}{\left(\frac{w_0}{10}\right)^3}\right]$$

$$-48\left[\frac{J_4\left(\frac{w_0}{2}\right)}{\left(\frac{w_0}{2}\right)^4} - \frac{J_4\left(\frac{w_0}{10}\right)}{\left(\frac{w_0}{10}\right)^4}\right]\right\} \qquad (2.49)$$

The image of a point-like object given for the conventional microscope is as follows:

$$I_{\text{1st model}}(r) = |h_{\text{1st model}}(r)|^2; \text{ for the 1st model} \qquad (2.50)$$

$$I_{\text{2nd model}}(r) = |h_{\text{2nd model}}(r)|^2; \text{ for the 2nd model} \qquad (2.51)$$

The image of a point in the confocal scanning laser microscope is given as follows:

$$I(r)_{\text{confocal}} = |h_{\text{1st model}}(r)|^4; \quad \text{for the 1st model} \tag{2.52}$$

$$I(r)_{\text{confocal}} = |h_{\text{2nd model}}(r)|^4; \quad \text{for the 2nd model} \tag{2.53}$$

The effective PSF in a confocal microscope is given as follows:

$$h_{\text{effective}}(r) = h_1(r)h_2(r) \tag{2.54}$$

h_1 represents the 1st lens, and h_2 represents the 2nd lens.

The plot of the absolute value of the PSF corresponding to the two models of hyper-resolving apertures at the horizontal line at 256 pixels is shown in Fig. 2.6. (a) The PSF corresponding to the aperture of quadratic-linear-constant segments arranged from the center with a ratio of 2:1:1, while (b) the PSF corresponding to the 2nd aperture of the dark disk ρ^6 distribution-linear-constant segments arranged from the center with a ratio of 0.1:0.4:0.25:0.25 for the radii.

The normalized absolute values of the PSF for the quadratic model in Fig. 2.7a, 1st model in Fig. 2.7b, and 2nd model in Fig. 2.7c are compared with those for a uniform circular aperture.

In Fig. 2.7, 2π NA (r_c/λ) 1st model $= 3.202 < 2\pi$ NA $(r_c/\lambda)_{\text{circular}} = 3.99$, as shown in (b), and 2π NA (r_c/λ) 2nd model $= 3.202 < 2\pi$ NA $(r_c/\lambda)_{\text{circular}} = 3.99$, as shown in (c), are compared with 2π NA $(r_c/\lambda)_{\text{quadratic}} = 3.202$, as shown in (a).

2.6 Hamming Aperture

The Hamming aperture is mathematically represented as follows:

$$P_{\text{ham}}(\rho) = [0.54 + 0.46\cos(\beta\pi(\rho))]; \quad 0 \le \rho \le \rho_0 \tag{2.55}$$

Then, the obstructed Hamming aperture is written as follows:

$$\begin{aligned} P_{\text{ham}}(\rho) &= [0.54 + 0.46\cos(\beta\pi(\rho - \rho_2))]; \quad \rho_2 \le \rho \le \rho_0 \\ &= 0; \quad 0 < \rho < \rho_2 \end{aligned} \tag{2.56}$$

ρ_2: the radius of the central obstructed disk, β is a parameter that has a value between zero and one.

This aperture represented by Eq. (2.56) is Fourier transformed to easily give this expression for the PSF [8, 9]:

$$h_{\text{ham}}(r) = F.T.\{P_{\text{ham}}(\rho)\} = F.T.[0.54 + 0.46\cos(\beta\pi(\rho - \rho_2))]$$

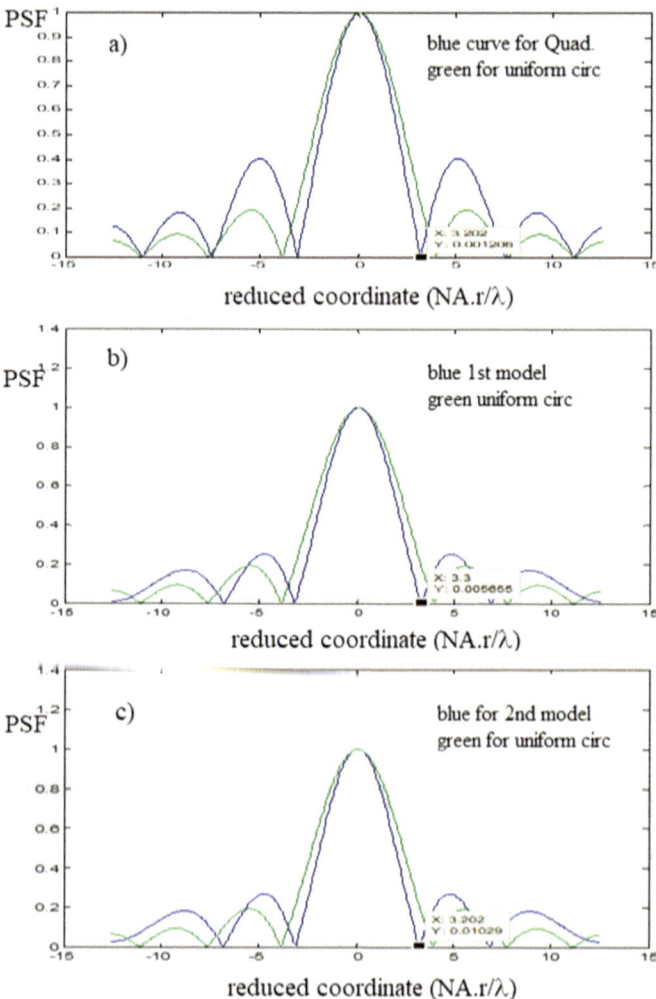

Fig. 2.7 The normalized PSF squared versus the reduced coordinate are shown in all the plotted Figures (a, b, and c). The comparative curve for the uniform circular aperture is green in color. The blue color is shown in all the plots. In a), for the quadratic aperture, in b) for the 1st model, and in c), for the 2nd model

$$= \left\{ \delta(r) + \left(\frac{1}{2} \right) \left[\delta_1 \left(r - \beta \, \lambda \frac{f}{2} - \beta \, \lambda \rho_2 \right) + \delta_2 \left(r + \beta \, \lambda \frac{f}{2} + \beta \, \lambda \rho_2 \right) \right] \right\}$$

$$(2.57)$$

Hence, the PSF of the Hamming aperture is computed to obtain a central Dirac delta function and two shifted Dirac delta functions. According to Eq. (2.57), the amount of shift is equal to $\pm \left(\beta \lambda \frac{f}{2} + \beta \lambda \rho_2 \right)$ from the center. Hence, the distance between the two shifted functions is $\beta\lambda(f + 2\rho_2)$ f: the focal length of the Fourier transform lens and λ : the wavelength of the laser radiation.

Referring to Eq. (2.56), the CTF, which is the convolution product of both Hamming apertures, is written in simple form as follows (Fig. 2.8):

$$CTF_{ham}(\rho) = [0.54 + 0.46\cos(\beta\pi(\rho - \rho_2))] \otimes [0.54 + 0.46\cos(\beta\pi(\rho - \rho_2))] \tag{2.58}$$

Referring to Figs. 2.9, and 2.10, the cutoff values of the PSF spatial frequencies are summarized as follows:

(a) The cutoff value $r_c = 2$ pixels for the Hamming aperture.
(b) The cutoff value $r_c = 3.8$ pixels for a uniform circular aperture

It is shown, from the comparison of the investigated apertures, that the Hamming aperture has a cutoff spatial frequency $r_c = 2$ pixels smaller than the cutoff corresponding to the uniform circular apertures of $r_c = 3.8$ pixels. Hence, the resolution of the Hamming aperture is better than that of the circular aperture.

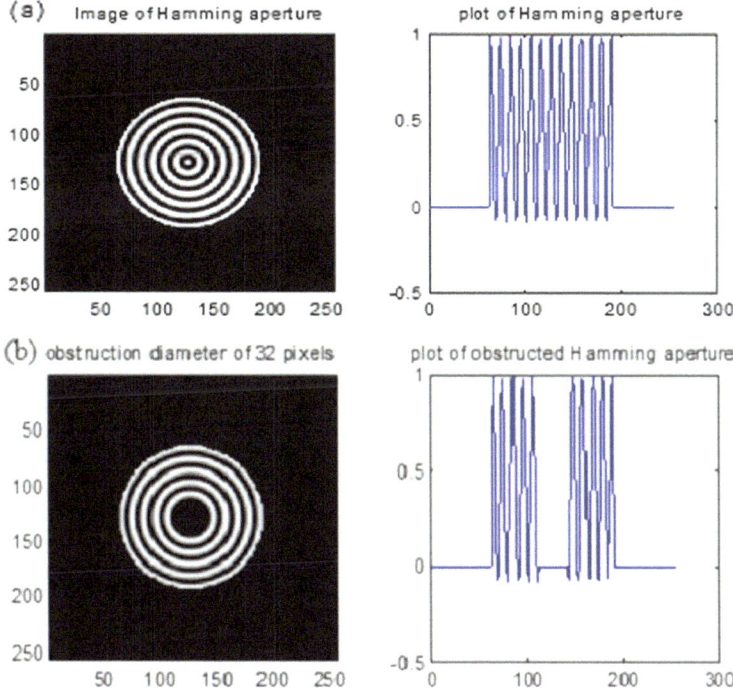

Fig. 2.8 An image of the Hamming aperture and its plot are shown in (**a**), while an image of the obstructed Hamming aperture and its plot are shown in (**b**). The total diameter is 128 pixels

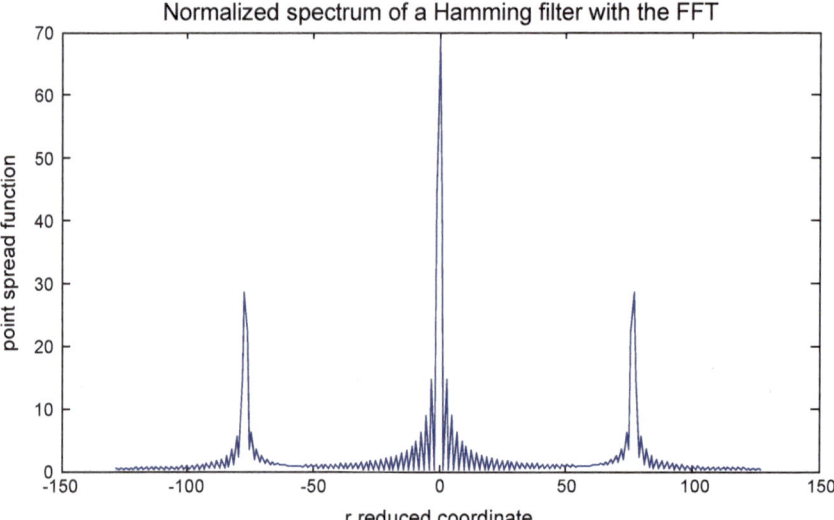

Fig. 2.9 The magnified PSF shape using a matrix of dimensions 256×256 pixels. Two symmetric side peaks are observed at 76 pixels around the central peak. The matrix dimensions are 256×256 pixels

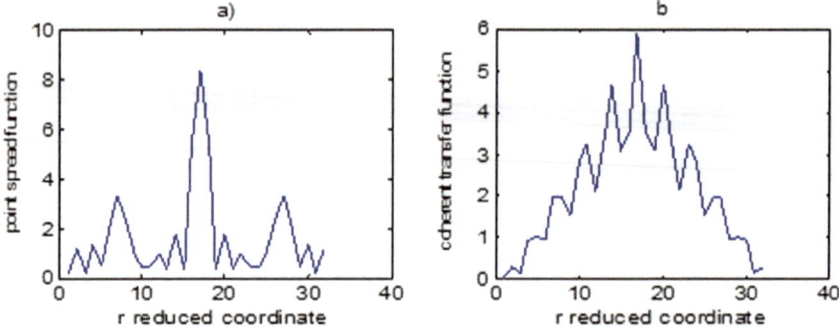

Fig. 2.10 For the Hamming aperture, the PSF, and the CTF are shown. The aperture has a total width of 16 pixels, and the CTF has a total width of 32 pixels. The total BW at FWHM $= 4$ pixels, and the cutoff spatial frequency $= 2$ pixels

2.7   Cauchy Aperture

The Cauchy aperture is mathematically represented as follows:

$$P_{\text{cauchy}}(\rho) = \frac{1}{1 + \beta^2 \left(\frac{\rho}{\rho_0}\right)^2}; \quad \left|\frac{\rho}{\rho_0}\right| \le 1 \text{ and } \beta < 1 \qquad (2.59)$$

β is a parameter, and ρ is the radial coordinate in the aperture plane with a maximum radius of ρ_0.

We represent the obstructed Cauchy aperture as the difference between the ordinary Cauchy aperture and a circular central zone of maximum radius $= \frac{\rho_0}{4}$ as follows:

$$P_{\text{Obst.}}(\rho) = P_{\text{Cauchy}}(\rho) - \text{circ}\left(\frac{4\rho}{\rho_0}\right) \tag{2.60}$$

We computed the PSF for the obstructed Cauchy aperture by performing the Fourier transform in polar coordinates as follows:

$$h(r) = \frac{1}{2\pi} \int_0^{2\pi} \int_{-\infty}^{\infty} P(\rho) \exp\left\{-j\frac{2\pi}{\lambda f}\rho\, r \cos(\theta)\right\} \rho\, d\rho\, d\theta \tag{2.61}$$

Symbolically, we write the PSF for the obstructed aperture as follows:

$$h(r) = \text{F.T.}\left\{P_{\text{Cauchy}}(\rho) - \text{circ}\left(\frac{4\rho}{\rho_0}\right)\right\} \tag{2.62}$$

Substituting Eq. (2.59) into Eq. (2.62), we write:

$$h(r) = \text{F.T.}\left[\frac{1}{1 + \beta^2\left(\frac{\rho}{\rho_0}\right)^2}\right] - \text{F.T.}\left[\text{circ}\left(\frac{4\rho}{\rho_0}\right)\right] = I_1(r) - I_2(r) \tag{2.63}$$

We solve the two transformations in Eq. (2.63) to obtain the following:

We solved the first integral in Eq. (2.63) by applying a Fourier transform, and we obtained the double-sided function represented in Eq. (2.64).

$$I_1(r) = \left(\frac{\pi}{\lambda f \beta}\right) \exp\left[-\frac{2\pi\rho_0}{\lambda f}|r|\right] \tag{2.64}$$

The double-sided function represented by Eq. (2.64) represents the PSF corresponding to the ordinary Cauchy aperture, where $r = r$; for $r > 0$, and $r = -r$; for $r < 0$.

The circular obstruction region has a PSF represented by an Airy disk as follows:

$$I_2(r) = \frac{2J_1(W_1)}{W_1}; \ W_1 = \frac{2\pi}{\lambda f}\left(\frac{\rho_0}{4}\right)r \tag{2.65}$$

Hence, from Eqs. (2.63), (2.64), and (2.65), we obtain the PSF corresponding to the obstructed Cauchy aperture as follows [10, 11]:

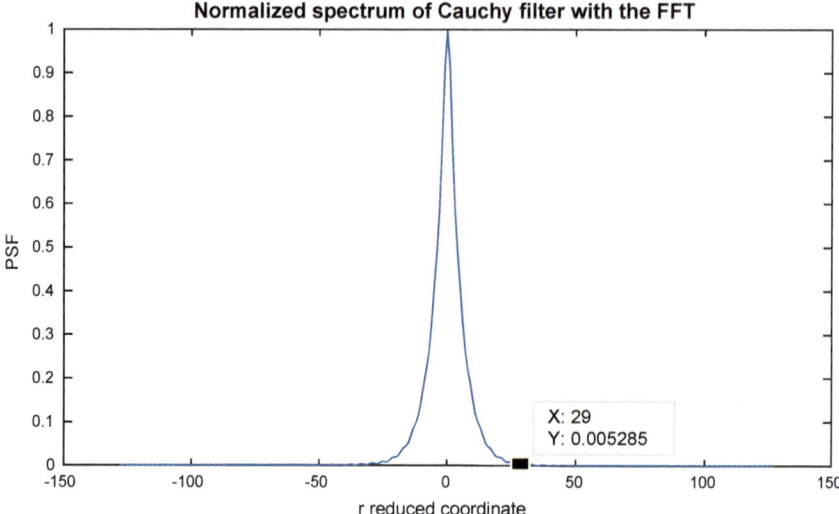

Fig. 2.11 Plot of the normalized PSF or the Cauchy Fourier spectrum corresponding to the aperture represented in Eq. (2.64). The aperture radius $= 64$ pixels

$$h(r) = \left(\frac{\pi}{\lambda f \beta}\right) \exp\left[-\frac{2\pi \rho_0}{\lambda f}|r|\right] - \frac{2J_1(W_1)}{W_1} \qquad (2.66)$$

We showed that the microscope resolution improved with the manipulation of obstructed Cauchy apertures. Hence, we obtain a compromise of resolution and contrast in the case of obstructed Cauchy apertures compared with the uniform circular apertures in the CSLM. Little degradation in image contrast is shown in the case of obstructed Cauchy apertures.

Additionally, we showed that legs in the PSF corresponded to the obstructed Cauchy aperture Fig. 2.11, compared with the ordinary Cauchy aperture Fig. 2.12, which may be useful for imaging extended objects (Fig. 2.13).

2.8 Polynomial Aperture

We propose five equal zones of higher-order polynomials ρ^8 at the center end with a linear function of ρ at the surface of the aperture as follows: ρ^8, ρ^6, ρ^4, ρ^2, and ρ.

The choice of five zones is presented to fulfill the arrangement assumed for the polynomial.

The assumed polynomial aperture has five equal zones of distribution, starting from the center, represented as shown in Fig. 2.14. The corresponding line plot is shown in the R.H.S. as in Fig. 2.14. In our case, the central zone has a transmission intensity proportional to ρ^8 instead of zero for the annular aperture.

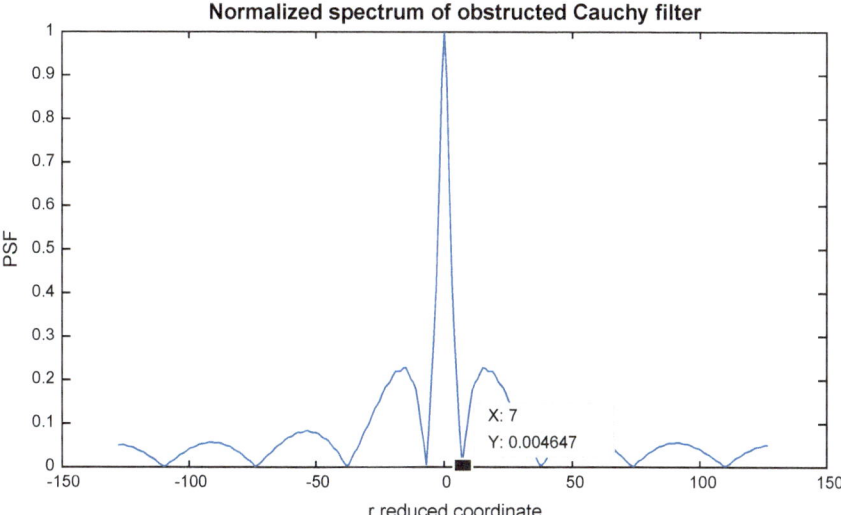

Fig. 2.12 Plot of the normalized PSF corresponding to the obstructed Cauchy aperture. The aperture radius = 64 pixels, and the obstruction central zone radius = 4 pixels

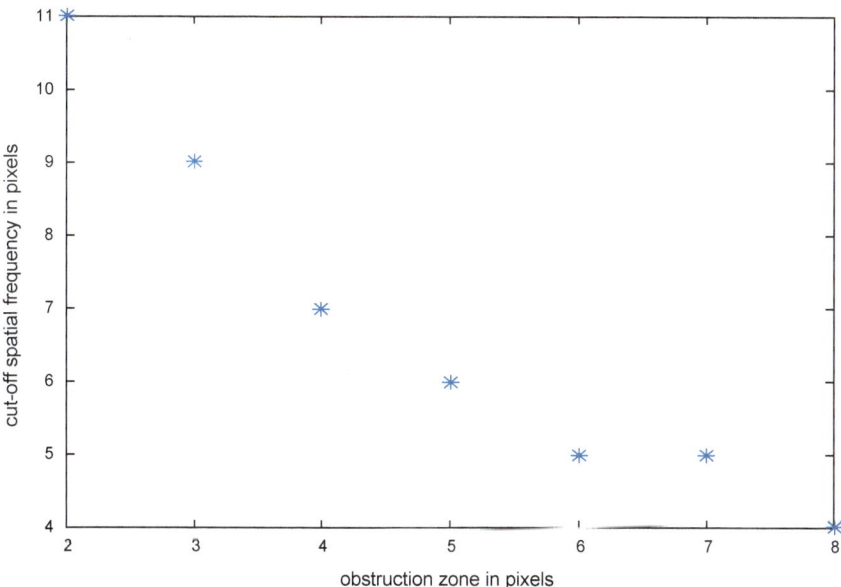

Fig. 2.13 The cutoff spatial frequency in pixels versus the obstruction central zone in pixels

Now, the polynomial aperture is written as follows:

$$
\begin{aligned}
P(\rho) &= a\rho^8, && \text{for } 0 \leq \rho < 0.2\rho_{max} \\
&= b\rho^6, && \text{for } 0.2\rho_{max} \leq \rho < 0.4\rho_{max} \\
&= c\rho^4, && \text{for } 0.4\rho_{max} \leq \rho < 0.6\rho_{max} \\
&= d\rho^2, && \text{for } 0.6\rho_{max} \leq \rho < 0.8\rho_{max} \\
&= e\rho, && \text{for } 0.8\rho_{max} \leq \rho < \rho_{max}
\end{aligned}
\tag{2.67}
$$

a, b, c, d, e, constants proportional to the cross-sectional areas of the corresponding zones.

In this model, referring to Eq. (2.67),

$$
a = 0.04\pi \ \rho_{max^2}, b = 0.12\pi \ \rho_{max^2}, c = 0.20\pi \ \rho_{max^2},
$$
$$
d = 0.28\pi \ \rho_{max^2}, e = 0.36\pi \ \rho_{max^2}
$$

Hence, $a + b + c + d + e = \pi \ \rho_{max}^2$ is the total area of the open circular aperture of radius ρ_{max}.

$\rho = (u, v)$ is the radial coordinate corresponding to the Cartesian coordinates (u, v), and ρ_{max} is the total aperture radius.

The PSF corresponds to the polynomial aperture, described in Eq. (2.67), computed by running the Fourier transform upon Eq. (2.67) considering coherent illumination emitted from the spatially filtered laser beam. Hence, the PSF is represented in integral form in polar coordinates as follows:

$$
h(r; \theta) = \int_0^{\rho_{max}} \int_0^{2\pi} P(\rho) \exp\left[-\frac{j2\pi}{\lambda f}\rho r \cos(\Phi - \theta)\right]\rho \, d\rho \, d\Phi
\tag{2.68}
$$

where $u = \rho \cos \Phi$, $v = \rho \sin \Phi$ are the Cartesian coordinates in the aperture plane corresponding to the polar coordinates (ρ, Φ), while $x = r \cos \theta$, $y = r \sin \theta$, are the Cartesian coordinates in the Fourier or focal plane corresponding to the polar coordinates (r, θ). The Fourier transform lens has a focal length $= f$.

Since the aperture has a circular symmetry of revolution, Eq. (2.68) is reduced to a function of r only as follows:

$$
h_{model\,1}(r) = 2\pi \int_0^{\rho_{max}} P(\rho) J_0\left(\frac{2\pi}{\lambda f}\rho r\right)\rho \, d\rho
\tag{2.69}
$$

$J_0(x)$ is the Bessel function of zero order, and the Bessel function of any order n $J_n(x)$ is represented by the following summation:

$$J_n(x) = \sum_{m=0}^{\infty} \frac{(-1)^m}{m!(m+n)!}\left(\frac{x}{2}\right)^{n+2m}.$$

By substituting Eq. (2.67) into Eq. (2.69), we obtain:

$$h_{\text{model1}}(r) = 2\pi \left[a \int_0^{0.2\rho_{\max}} \rho^8 J_0\left(\frac{2\pi}{\lambda f}\rho r\right)\rho d\rho + b \int_{0.2\rho_{\max}}^{0.4\rho_{\max}} \rho^6 J_0\left(\frac{2\pi}{\lambda f}\rho r\right)\rho d\rho \right.$$

$$+ c \int_{0.4\rho_{\max}}^{0.6\rho_{\max}} \rho^4 J_0\left(\frac{2\pi}{\lambda f}\rho r\right)\rho d\rho + d \int_{0.6\rho_{\max}}^{0.8\rho_{\max}} \rho^2 J_0\left(\frac{2\pi}{\lambda f}\rho r\right)\rho d\rho$$

$$+ e \int_{0.8\rho_{\max}}^{\rho_{\max}} \rho J_0\left(\frac{2\pi}{\lambda f}\rho r\right)\rho d\rho] \tag{2.70}$$

By solving Eq. (2.70), we obtain the corresponding result for the PSF as follows [11]:

$$h_{\text{model 1}}(r) = \frac{J_1(W_5)}{W_5} - 0.08\sum_{i=1}^{4}\frac{J_1(W_i)}{W_i} + 0.4\frac{J_2(W_1)}{W_1^2} + 0.08\frac{J_2(W_2)}{W_2^2} - 0.24\frac{J_2(W_3)}{W_3^2}$$

$$- 0.56\frac{J_2(W_4)}{W_4^2} + 0.36\left(\frac{J_0(W_5)}{W_5^2} - \frac{J_0(W_4)}{W_4^2}\right) - 0.96\frac{J_3(W_1)}{W_1^3} + 1.28\frac{J_3(W_2)}{W_2^3}$$

$$+ 1.6\frac{J_3(W_3)}{W_3^3} + 0.72\sum_{i=1}^{N}\left(\frac{J_i(W_4)}{W_4^3} - \frac{J_i(W_5)}{W_5^3}\right) - 1.92\frac{J_4(W_1)}{W_1^4} - 5.76\frac{J_4(W_2)}{W_2^4}$$

$$+ 15.36\frac{J_5(W_1)}{W_1^5} \tag{2.71}$$

where $i = (1, 3, 5, \ldots, N)$. $W_1 = \frac{2\pi}{\lambda f}(0.2\rho_{\max})r$, $W_2 = \frac{2\pi}{\lambda f}(0.4\rho_{\max})r$, $W_3 = \frac{2\pi}{\lambda f}(0.6\,\rho_{\max})r$; $W_4 = \frac{2\pi}{\lambda f}(0.8\,\rho_{\max})r$, $W_5 = \frac{2\pi}{\lambda f}(\rho_{\max})$.

According to the numerical results that were obtained via the FFT technique, the polynomial aperture yields a better PSF curve of the spatial frequency cutoff than the uniform circular and linear apertures (Fig. 2.15).

$$W_{\text{cutoff}} = 0.81 \text{ (polynomial)} < W_{\text{cutoff}} = 0.86\text{(linear)} < W_{\text{cutoff}} = 1.0 \text{ (circular)}$$

The PSF corresponds to the 1st model using the analytical solution represented by Eq. (2.71) is compared with a uniform circular aperture. In the computation, it is assumed that $\lambda = 500$ nm and the NA $= 0.5$. It shows that:

$$W_{\text{cutoff}} = 4.417 \text{ (quadratic)} < W_{\text{cutoff}} = 4.795\text{(polynomial)} < W_{\text{cutoff}}$$
$$= 5.097 \text{ (circular)}$$

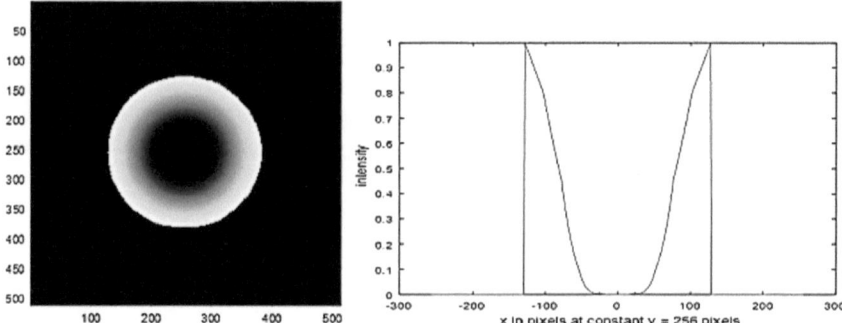

Fig. 2.14 In the L.H.S., a grayscale image of a circular aperture in the form of a polynomial distribution with five equal zones. The concentric zones have distributions ρ^8, ρ^6, ρ^4, ρ^2, and ρ computed from the aperture center. The matrix dimensions are 512×512 pixels, and the total radius of the aperture $= 128$ pixels. In the R.H.S., the intensity plot of the polynomial aperture at the center of the aperture at constant $y = 256$ pixels

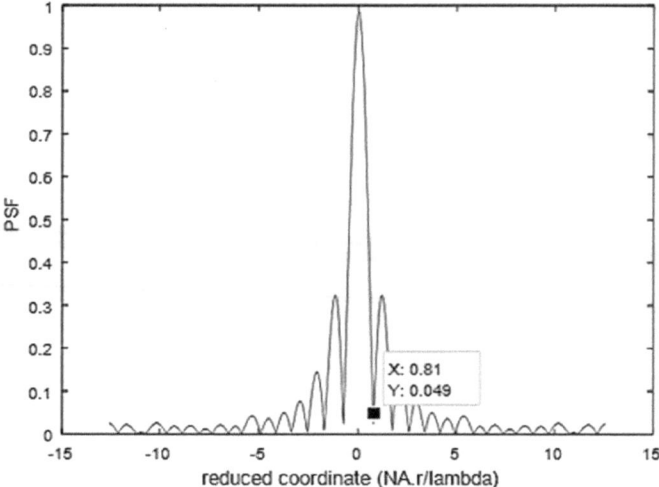

Fig. 2.15 The normalized PSF for the 1st model of the polynomial aperture using the FFT technique. The total diameter $= 32$ pixels, and the cutoff spatial frequency is found at a $W_{cutoff} = 0.81$

2.9   New Model of the Modulated Aperture [12]

A design of eight equally spaced circles placed at equal distances from the origin is suggested. Three models, corresponding to the eight-circle design considering conic, linear, and quadratic distributions were investigated. This arrangement was considered for the sake of improving both the microscope resolution and image

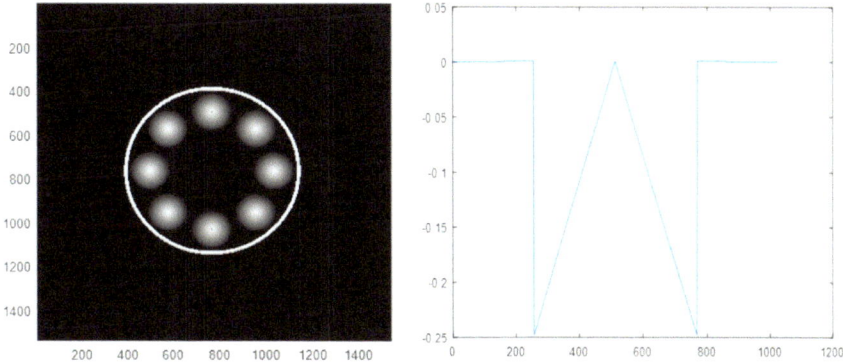

Fig. 2.16 Eight conic apertures placed tangent to the annular aperture. The annular width $= 16$ pixels, and the external radius $= 350$ pixels. The radius of each conical aperture $= 64$ pixels, and the plot of the conical aperture with a radius $= 256$ pixels placed at 512,512 pixels.s

contrast compared with those of the pure annular aperture. This design is different from that used in other recent work on aperture modulation.

The eight equally spaced conic apertures shown in Fig. 2.16 are described as follows:

$$P_T(x, y) = P_1 + P_2 + \cdots + P_8 \tag{2.72}$$

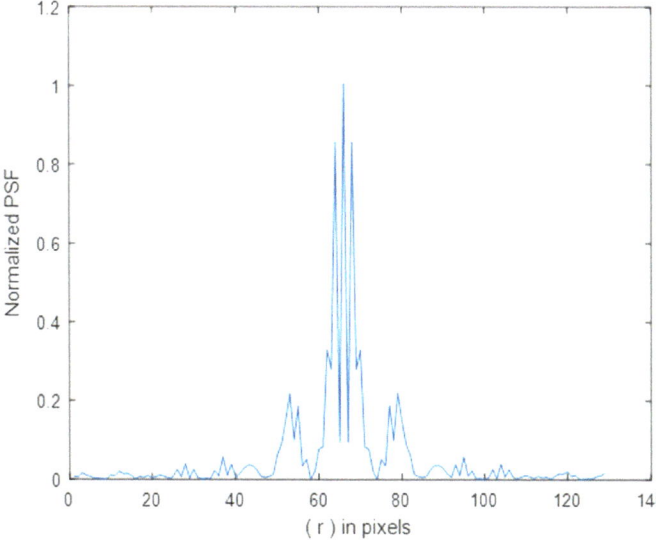

Fig. 2.17 Normalized PSF as a function of the radial distance r (pixels) in the Fourier plane for a lens limited by the aperture shown in Fig. (2.16)

with the apertures $P_{1,2}$ found along the x-axis at distances x_0 and the apertures $P_{3,4}$ found along the y-axis at distances y_0 represented as follows:

$$P_{1,2}(x, y) = 1 - \sqrt{(x \pm x_0)^2 + y^2} \tag{2.73}$$

$$P_{3,4}(x, y) = 1 - \sqrt{x^2 + (y \pm y_0)^2} \tag{2.74}$$

For rotated coordinates by an angle of 45°, another four apertures are shown in the same Fig. 2.16 and are represented as follows:

$$P_{5,6}(x, y) = 1 - \sqrt{\left(x \pm \frac{x_0}{\sqrt{2}}\right)^2 + \left(y \pm \frac{y_0}{\sqrt{2}}\right)^2} \tag{2.75}$$

$$P_7(x, y) = 1 - \sqrt{\left(x - \frac{x_0}{\sqrt{2}}\right)^2 + \left(y + \frac{y_0}{\sqrt{2}}\right)^2} \tag{2.76}$$

$$P_8(x, y) = 1 - \sqrt{\left(x + \frac{x_0}{\sqrt{2}}\right)^2 + \left(y - \frac{y_0}{\sqrt{2}}\right)^2} \tag{2.77}$$

The Fourier transform corresponding to this model of eight conical apertures is not easy to solve, but we use the FFT technique to solve it by obtaining the PSF.

Similarly, the PSF is computed for the other eight equally spaced apertures of the linear and quadratic distributions using the FFT technique.

The computation of the PSF using the FFT for all apertures. The normalized PSF as a function of the radial distance r (pixels) in the Fourier plane for a lens limited by the aperture in Fig. 2.16 is shown in Fig. 2.17.

2.10 Elliptic Apertures

The studied aperture is assumed to be elliptical:

$$p(x, y) = 1; \quad \frac{x^2}{a^2} + \frac{y^2}{b^2} \leq 1$$
$$= 0; \quad \frac{x^2}{a^2} + \frac{y^2}{b^2} > 1 \tag{2.78}$$

The point spread function of this elliptical aperture is computed numerically by running the Fourier transform on Eq. (2.78). This transformation is written in integral form as follows:

$$h(u, v) = \int\!\!\int_{-\infty}^{\infty} p(x, y)e^{-\left(\frac{2\pi i}{\lambda f}\right)(xu+yv)}\, dx\, dy \tag{2.79}$$

We stand for the elliptical aperture as an image of matrix dimensions $N \times M$ pixels where $N = M$ and run the two-dimensional F.T. on this numerical aperture to obtain the point spread function $h(u, v)$ and its intensity impulse response as:

$$h(u, v) = \sum_{n=1}^{N} \sum_{m=1}^{M} p(m\triangle x, n\triangle y)e^{-\left(\frac{2\pi i}{\lambda f}\right)(m\triangle x u + n\triangle y v)}\triangle x\, \triangle y \tag{2.80}$$

$$I(u, v) = |h(u, v)|^2 \tag{2.81}$$

Returning to the transformation integral in Eq. (2.79), we can solve it numerically as follows:

From the equation for the elliptical aperture $x^2/a^2 + y^2/b^2 = 1$, we obtain x in terms of y as:

$$x = a\sqrt{1 - \left(\frac{y}{b}\right)^2} \tag{2.82}$$

Substituting Eq. (2.82) into Eq. (2.79) and using Eq. (2.78), the double integral is transformed into a single integral in the variable y as follows:

$$h(u) = \frac{a}{b} \int_{b}^{-b} \exp\left\{-\frac{j2\pi}{\lambda f}\left[au\sqrt{1 - \left(\frac{y}{b}\right)^2} + \sqrt{r^2 - u^2}y\right]\right\}\left(\frac{y \cdot dy}{\sqrt{b^2 - y^2}}\right) dy \tag{2.83}$$

This integral is solved numerically by transforming it into a summation as follows [13]:

$$h(u) = \frac{2a}{b}\sum_{m=0}^{M}\frac{m}{\sqrt{b^2 - (m\triangle y)^2}}\exp\left\{-\frac{j2\pi}{\lambda f}\left[au\sqrt{1 - \left(\frac{m\triangle y}{b}\right)^2}\right.\right.$$
$$\left.\left. +\sqrt{r^2 - u^2}m\triangle y\right]\right\}(\triangle y)^3 \tag{2.84}$$

where m is an integer that changes from zero to M and $r = \sqrt{u^2 + v^2}$ is the polar coordinate assumed in the Fourier plane.

The point spread function for the elliptical aperture is plotted in Fig. 2.18b, where the semi major axis is $a = 1$ and the semi-minor axis is $b = 0.5$. The PSF was distinguished compared with the circular aperture Fig. 2.18a, by the lower height of the 1st side lobe compared with the height of the 2nd lobe.

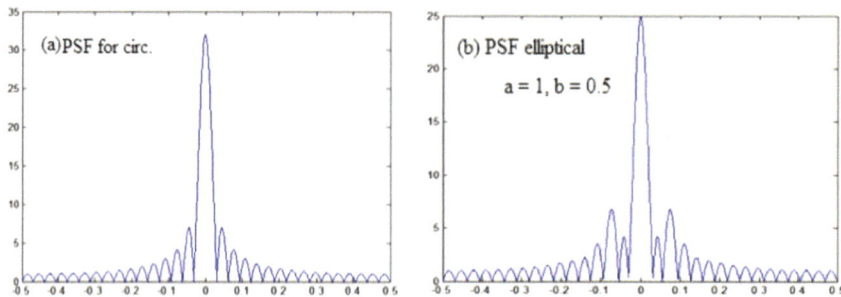

Fig. 2.18 PSF corresponding to circular apertures in (**a**) and elliptical apertures in (**b**)

2.11   Rectangular Apertures

A rectangular aperture composed of four equal squares was placed at equal distances from the center. The total matrix of rectangular shapes has dimensions of 1024 × 1024 pixels. The dimensions of each square $(x_0, y_0) = 128 \times 128$ pixels are located along the Cartesian coordinates (x, y) at distances $\pm x_d$ and $\pm y_d$, as shown in Fig. 2.19a. A central obstruction is governed by the difference between two circles where the width of the annulus = 32 pixels. The uniform illumination emitted from the laser beam is incident upon this new aperture, hence, the amplitude transmittance is written as follows:

$$A(x, y) = \text{rect}(x - x_d, y) + \text{rect}(x + x_d, y) + \text{rect}(x, y - y_d) +$$
$$\text{rect}(x, y + y_d) + \text{circ}(r_1) - \text{circ}(r_2) \tag{2.85}$$

where

$$\text{rect}(x, y) = 1; \; |\frac{x}{x_0}| \le 1, |\frac{y}{y_0}| \le 1$$

$$\text{circ}(r_1) = 1; \; |\frac{r}{r_{01}}| \le 1, \; \text{and circ}(r_2) = 1; \; |\frac{r}{r_{02}}| \le 1$$

The PSF corresponding to the novel aperture described by Eq. (2.85) is obtained by performing the Fourier transform as follows:

$$\begin{aligned}
h(x, y) &= \text{F.T.} \{A(x, y)\} = \text{F.T.} \{\text{rect}(x - x_d, y) \\
&+ \text{rect}(x + x_d, y) + \text{rect}(x, y - y_d) \\
&+ \text{rect}(x, y + y_d) + \text{circ}(r_1) - \text{circ}(r_2)\} \\
&= \text{F.T.} \{\text{rect}(x, y) \otimes \delta(x - x_d, y)\} + \text{F.T.} \{\text{rect}(x, y) \otimes \delta (x + x_d, y)\} \\
&+ \text{F.T.} \{\text{rect}(x, y) \otimes \delta (x, y - y_d)\} + \text{F.T.} \{\text{rect}(x, y) \otimes \delta(x, y + y_d)\} \\
&+ \text{F.T.} \{\text{circ}(r_1)\} - \text{F.T.} \{\text{circ}(r_2)\} \tag{2.86}
\end{aligned}$$

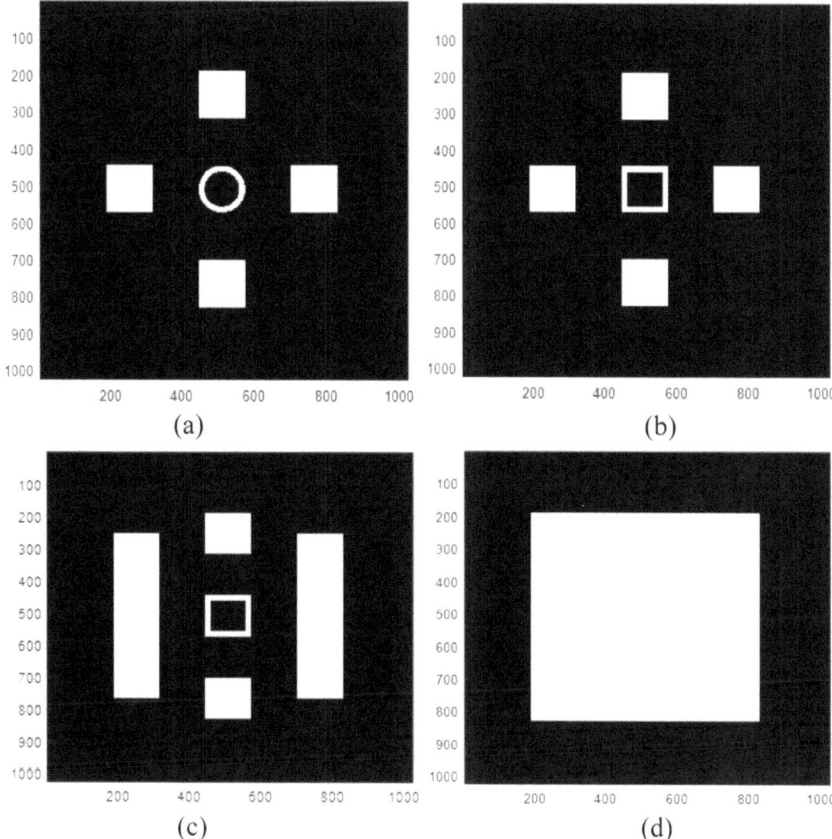

Fig. 2.19 In Fig. 2.19a, an aperture composed of four squares arranged symmetrically along the Cartesian coordinates and obstructed by a circular annulus placed in the center of 128-pixel diameter and 16-pixel annular width is shown. Each square has dimensions of 128×128 pixels and is shifted by 256 pixels along the coordinates. In Fig. 2.19b, the aperture is like that shown in Fig. 2.19a, except that the circular annulus is replaced by a rectangular annulus. In Fig. 2.19c, an aperture is composed of two squares arranged symmetrically along the y-axis and two rectangles arranged along the x-axis and obstructed by an annular rectangle placed in the center. In Fig. 2.19d, an aperture composed of a square with dimensions of 640×640 pixels is shown

The 1st transformation is solved considering that the convolution of two functions is equivalent to the simple product of the F.T. of each function, hence, we write:

F.T.$\{\text{rect}(x, y) \otimes \delta(x - x_d, y)\} = $ F.T.$\{\text{rect}(x, y)\}.$F.T.$\{(x - x_d, y)\}.$

It is known that the Fourier transform of the rect function is written as:

$$\text{F.T.}\{\text{rect}(x, y)\} = x_0 y_0 \left[\frac{\sin\left(\frac{\pi x_0 u}{\lambda f}\right)}{\left(\frac{\pi x_0 u}{\lambda f}\right)} \right] \left[\frac{\sin\left(\frac{\pi y_0 v}{\lambda f}\right)}{\left(\frac{\pi y_0 v}{\lambda f}\right)} \right] = x_0 y_0 sinc(x\prime)sinc(y\prime) \quad (2.87)$$

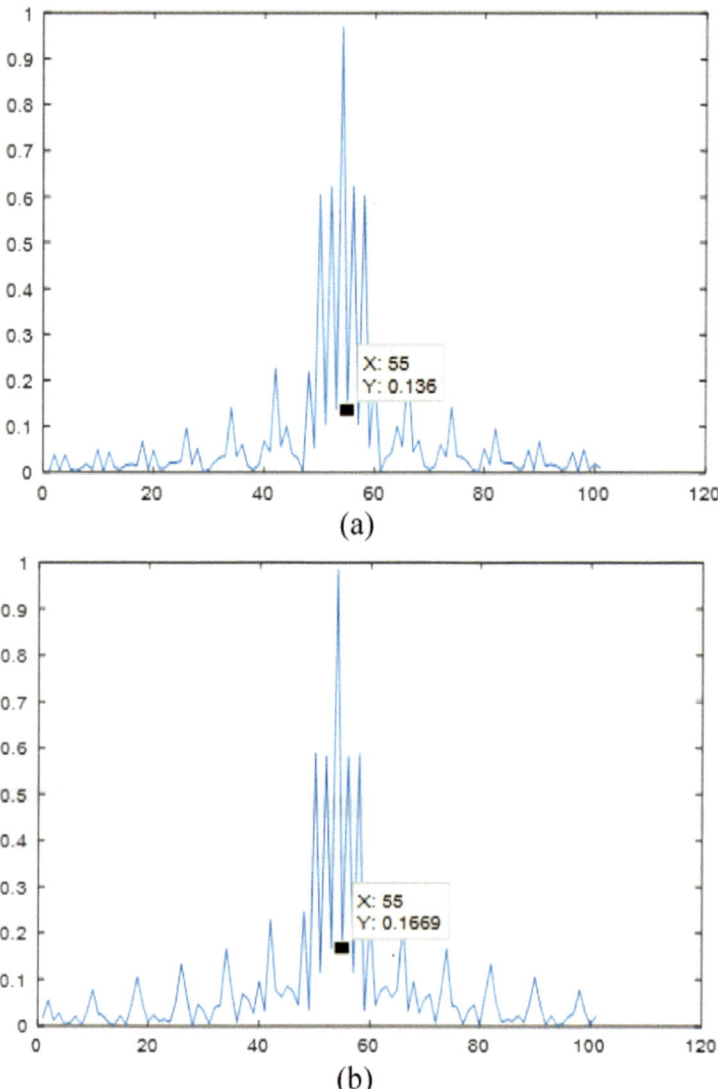

(a)

(b)

Fig. 2.20 The PSF for the aperture shown in Fig. 2.20a, in the range of 460–560 pixels. The cutoff for the central lope is = 55 pixels, while the central peak is 54 pixels. FWHM = 1 pixel. Figure 2.20bThe PSF for the aperture shown in Fig. 2.19b, in the range of 460–560 pixels. The cutoff for the central lope is = 55 pixels, while the central peak is 54 pixels. FWHM = 1 pixel. Figure 2.20c: The PSF for the aperture shown in Fig. 2.19c, in the range of 460–560 pixels. The cutoff for the central lope is = 55 pixels, while the central peak is 54 pixels. FWHM = 1 pixel. The side lobes are strengthened compared with those shown in Fig. 2.20b. Figure 2.20d: The PSF for the whole aperture shown in Fig. 2.20d, is in the same range as that in Fig. 2.20b. The cutoff for the central lope is = 57 pixels, while the central peak is 54 pixels. FWHM = 3 pixels

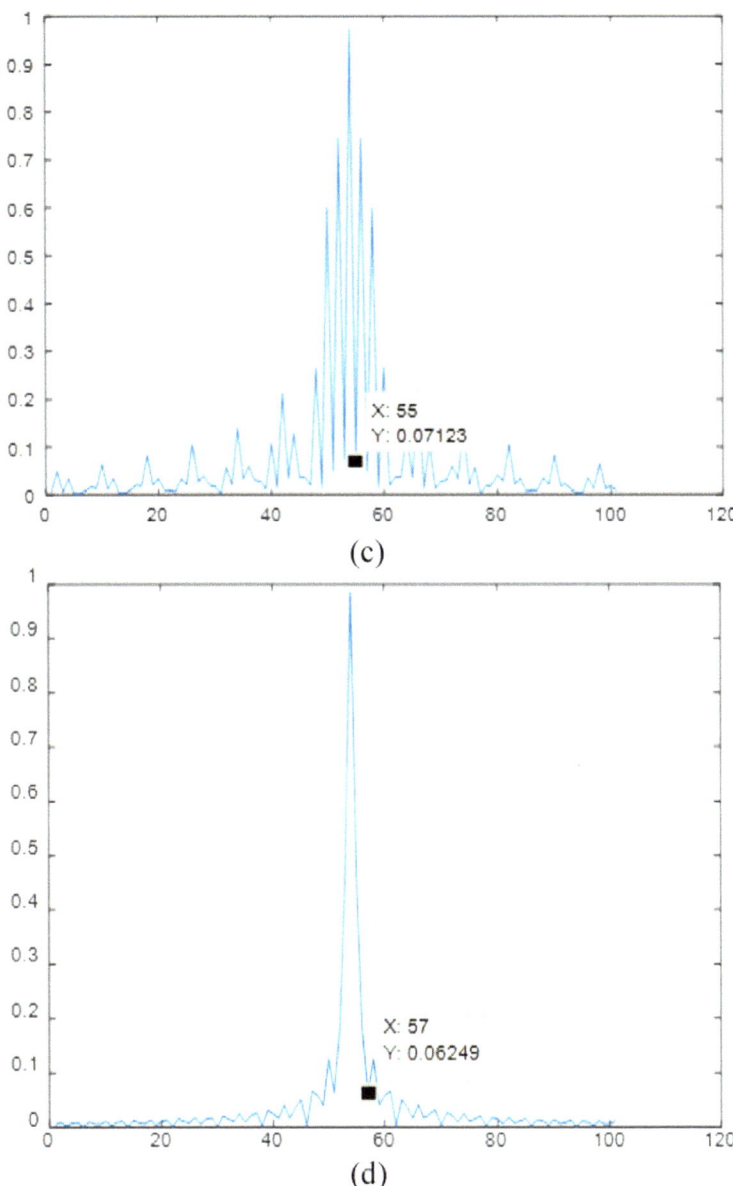

Fig. 2.20   (continued)

where the sinc function is defined as:

$\text{sinc}(x') = \sin(\pi x')/\pi x'$, $x' = x_0\, u/\lambda f$, and a similar expression for $\text{sinc}(y')$.

Additionally, the Fourier transform of the circ function is obtained as follows [11]:

$$\text{F.T.}\{\text{circ}(r_1)\} = 2J_1(Z_1)/Z_1, \text{ and F.T.}\{\text{circ}(r_2)\} = 2J_1(Z_2)/Z_2$$

where $Z_1 = 2\pi r_{01}\frac{w_1}{\lambda f}$ and $Z_2 = 2\pi r_{02}\frac{w_2}{\lambda f}$ represent the reduced coordinates in the Fourier planes corresponding to the external and internal circles of radii r_{01} and r_{02}, respectively. $w_{1,2} = \sqrt{u^2 + v^2}$ is the radial coordinate in the Fourier plane corresponding to each circle.

The Fourier transform of the shifted Dirac delta function gives an inclined plane wave as:

$$\text{F.T.}\{\delta(x - x_d, y)\}$$

$$= \int\limits_{-\infty}^{\infty}\int \delta(x - x_d, y)\exp\left[-\frac{j2\pi}{\lambda f}(xu + yv)\right]dxdy$$

$$= \exp\left(-\frac{j2\pi x_d u}{\lambda f}\right)\int\limits_{-\infty}^{\infty}\int \delta(x - x_d, y)\exp\left[-\frac{j2\pi}{\lambda f}((x - x_d)u + yv)\right]dxdy$$

$$= \exp\left(-\frac{j2\pi x_d u}{\lambda f}\right) \tag{2.88}$$

Similar expressions are obtained for the other shifts of the Dirac delta function as follows:

$$\text{F.T.}\{\delta(x + x_d, y)\} = \exp\left(+\frac{j2\pi x_d u}{\lambda f}\right) \tag{2.89}$$

$$\text{F.T.}\{\delta(x, y - y_d)\} = \exp\left(-\frac{j2\pi y_d v}{\lambda f}\right) \tag{2.90}$$

$$\text{F.T.}\{\delta(x, y + y_d)\} = \exp\left(+\frac{j2\pi y_d v}{\lambda f}\right) \tag{2.91}$$

Plugging Eqs. (2.87–2.91) in Eq. (2.86), we obtain:

$$h(x, y) = x_0 y_0 \text{sinc}(x')\text{sinc}(y')\left\{\exp\left(-\frac{j2\pi x_d u}{\lambda f}\right) + \exp\left(+\frac{j2\pi x_d u}{\lambda f}\right) + \exp\left(-\frac{j2\pi y_d v}{\lambda f}\right)\right.$$

$$\left. + \exp\left(+\frac{j2\pi y_d v}{\lambda f}\right)\right\} + \left\{\frac{2J_1(Z_1)}{Z_1} - \frac{2J_1(Z_2)}{Z_2}\right\} \tag{2.92}$$

Hence, the PSF is finally written as follows [14]:

$$h(x, y) = 2x_0y_0\text{sinc}(x')\text{sinc}(y')\left\{\cos\left(\frac{2\pi x_d u}{\lambda f}\right) + \cos\left(\frac{2\pi y_d v}{\lambda f}\right)\right\}$$
$$+ \left\{\frac{2J_1(Z_1)}{Z_1} - \frac{2J_1(Z_2)}{Z_2}\right\} \tag{2.93}$$

Equation (2.93) is rewritten as follows:

$$h(x, y) = 4x_0y_0\text{sinc}(x')\text{sinc}(y')\left\{\cos\left(\frac{\pi(x_d u + y_d v)}{\lambda f}\right)\cos\left(\frac{\pi(x_d u - y_d v)}{\lambda f}\right)\right\}$$
$$+ \left\{\frac{2J_1(Z_1)}{Z_1} - \frac{2J_1(Z_2)}{Z_2}\right\} \tag{2.94}$$

The intensity impulse response is computed from the modulus square of Eq. (2.94) as follows:

$$I(x, y)$$
$$= 16x_0^2 y_0^2 \text{sinc}^2(x')\text{sinc}^2(y')\left\{\cos^2\left(\frac{\pi(x_d u + y_d v)}{\lambda f}\right)\cos^2\left(\frac{\pi(x_d u - y_d v)}{\lambda f}\right)\right\}$$
$$+ 8x_0y_0\text{sinc}(x')\text{sinc}(y')\left\{\cos\left(\frac{\pi(x_d u + y_d v)}{\lambda f}\right)\cos\left(\frac{\pi(x_d u - y_d v)}{\lambda f}\right)\right\}$$
$$\left\{\frac{2J_1(Z_1)}{Z_1} - \frac{2J_1(Z_2)}{Z_2}\right\} + \left\{\frac{2J_1(Z_1)}{Z_1} - \frac{2J_1(Z_2)}{Z_2}\right\}^2 \tag{2.95}$$

The apertures and the corresponding PSF results are represented in Figs. 2.20 and 2.21.

2.12 PSF of Hexagonal Apertures

We summarize the PSF results for the recently presented hexagonal aperture [15] as follows:

$$P(x, y) = \text{rect}(x, y) + \text{tri}\left(x, y - \frac{2a}{3}\right) + \text{tri}\left(x, y + \frac{2a}{3}\right) \tag{2.96}$$

where $\text{rect}(x, y) = 1$; $|\frac{x}{b}| \leq 1$ and $|\frac{y}{b}| \leq 1$ represents a rectangle of sides a and b, where $a = b$.

$\text{tri}\left(x, y - \frac{2a}{3}\right) = 1$. It represents a triangle of base $2b$ and height a.

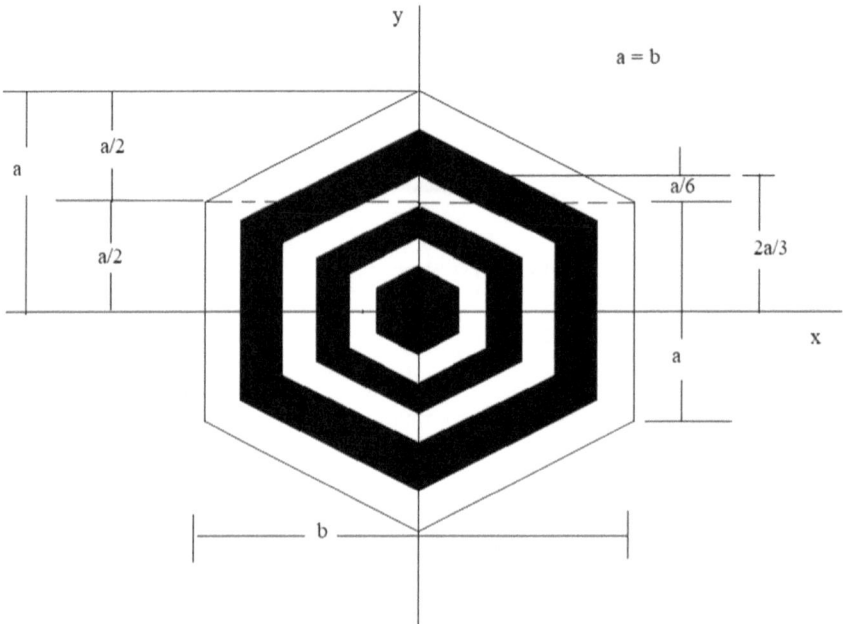

Fig. 2.21 A concentric hexagonal black and white pupil where the number of zones is $N = 6$ and the outer side length of the hexagon is a

We applied the Fourier transform using Eq. (2.96), and we obtained the PSF as follows [15]:

$$h(u, v) = \left[\frac{\sin(\pi bu)}{\pi bu}\right]\left[\frac{\sin(\pi av)}{\pi av}\right]\left\{1 + 2\left[\frac{\sin(\pi bu)}{\pi bu}\right]\left[\frac{\sin(\pi av)}{\pi av}\right]\cos\left[\left(\frac{4\pi a}{3\lambda f}\right)v\right]\right\} \quad (2.97)$$

The intensity corresponding to the PSF is written as follows:

$$I(u, v) = |h(u, v)|^2 \quad (2.98)$$

In this chapter, we compute the PSF corresponding to six concentric zones of black and white hexagonal shapes, as shown in Fig. (2.23).

We represent the B/W hexagonal pupil as follows:

$$P(x, y) = \left\{\text{rect}(x, y) + \text{tri}\left(x, y - \frac{2a}{3}\right) + \text{tri}\left(x, y + \frac{2a}{3}\right)\right\}$$
$$- \left\{\text{rect}(\alpha_1 x, \beta_1 y) + \text{tri}\left(\alpha_1 x, y - \frac{a}{3}\right) + \text{tri}\left(\alpha_1 x, y + \frac{a}{3}\right)\right\}$$
$$+ \left\{\text{rect}(\alpha_2 x, \beta_2 y) + \text{tri}\left(\alpha_2 x, y - \frac{2a}{9}\right) + \text{tri}\left(\alpha_2 x, y + \frac{2a}{9}\right)\right\}$$

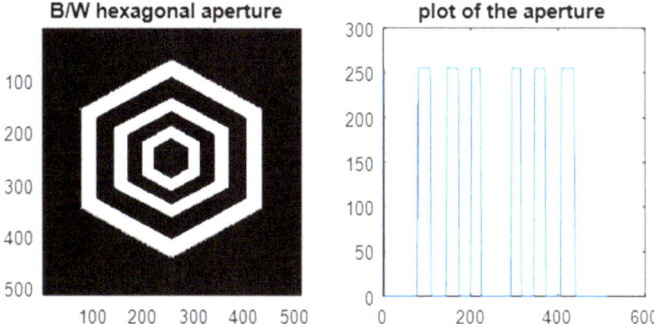

Fig. 2.22 The image is a B/W hexagonal pupil where the number of zones $N = 6$ and the corresponding line plot at $y = 256$ pixels

$$-\left\{ \text{rect}(\alpha_3 x, \beta_3 y) + \text{tri}\left(\alpha_3 x, y - \frac{2a}{9}\right) + \text{tri}\left(\alpha_3 x, y + \frac{2a}{9}\right) \right\}$$

$$+\left\{ \text{rect}(\alpha_4 x, \beta_4 y) + \text{tri}\left(\alpha_4 x, y - \frac{2a}{9}\right) + \text{tri}\left(\alpha_4 x, y + \frac{2a}{9}\right) \right\}$$

$$-\left\{ \text{rect}(\alpha_5 x, \beta_5 y) + \text{tri}\left(\alpha_5 x, y - \frac{2a}{9}\right) + \text{tri}\left(\alpha_5 x, y + \frac{2a}{9}\right) \right\} \quad (2.99)$$

where $\alpha_1 = \dfrac{a}{2}, \alpha_2 = \dfrac{a}{3}, \alpha_3 = \dfrac{a}{4}, \alpha_4 = \dfrac{a}{5}, \alpha_5 = \dfrac{a}{6},$ and $\alpha = \beta.$

For any definite number N of hexagonal zones, we wrote the aperture as follows:

$$P(x, y) = \sum_{i=1}^{N}\left\{ \text{rect}(\alpha_i x, \beta_i y) + \text{tri}\left(\alpha_i x, y - \frac{2a}{(3 \times i)}\right) + \text{tri}\left(\alpha_i x, y + \frac{2a}{(3 \times i)}\right) \right\}$$

$$-\sum_{j=2}^{N}\left\{ \text{rect}(\alpha_j x, \beta_j y) + \text{tri}\left(\alpha_j x, y - \frac{2a}{(3 \times j)}\right) + \text{tri}\left(\alpha_j x, y + \frac{2a}{(3 \times j)}\right) \right\}$$

$$(2.100)$$

Where $i = 1, 3, 5, \ldots$ is an odd number and $j = 2, 4, 6, \ldots$ even number.

We apply the Fourier transform to the modulated hexagonal aperture represented in Eq. (2.100), and we obtain the PSF as follows [16]:

$$h(u, v) = \left[\frac{\sin(\pi bu)}{\pi bu}\right]\left[\frac{\sin(\pi av)}{\pi av}\right] \times \left\{1 + 2\left[\frac{\sin(\pi bu)}{\pi bu}\right]\left[\frac{\sin(\pi av)}{\pi av}\right]\cos\left[\left(\frac{4\pi a}{3\lambda f}\right)v\right]\right\}$$

$$-\left[\frac{\sin(\pi\alpha_1 bu)}{\pi\alpha_1 bu}\right]\left[\frac{\sin(\pi\beta_1 av)}{\pi\beta_1 av}\right] \times \left\{1 + 2\left[\frac{\sin(\pi\alpha_1 bu)}{\pi\alpha_1 bu}\right]\left[\frac{\sin(\pi\beta_1 av)}{\pi\beta_1 av}\right]\cos\left[\left(\frac{2\pi}{\lambda f}\right)\alpha_1 v\right]\right\}$$

$$+\left[\frac{\sin(\pi\alpha_2 bu)}{\pi\alpha_2 bu}\right]\left[\frac{\sin(\pi\beta_2 av)}{\pi\beta_2 av}\right] \times \left\{1 + 2\left[\frac{\sin(\pi\alpha_2 bu)}{\pi\alpha_2 bu}\right]\left[\frac{\sin(\pi\beta_2 av)}{\pi\beta_2 av}\right]\cos\left[\left(\frac{2\pi a}{\lambda f}\right)\alpha_2 v\right]\right\}$$

$$-\left[\frac{\sin(\pi\alpha_3 bu)}{\pi\alpha_3 bu}\right]\left[\frac{\sin(\pi\beta_3 av)}{\pi\beta_3 av}\right] \times \left\{1 + 2\left[\frac{\sin(\pi\alpha_3 bu)}{\pi\alpha_3 bu}\right]\left[\frac{\sin(\pi\beta_3 av)}{\pi\beta_3 av}\right]\cos\left[\left(\frac{2\pi}{\lambda f}\right)\alpha_3 v\right]\right\}$$

$$+\left[\frac{\sin(\pi\alpha_4 bu)}{\pi\alpha_4 bu}\right]\left[\frac{\sin(\pi\beta_4 av)}{\pi\beta_4 av}\right] \times \left\{1 + 2\left[\frac{\sin(\pi\alpha_4 bu)}{\pi\alpha_4 bu}\right]\left[\frac{\sin(\pi\beta_4 av)}{\pi\beta_4 av}\right]\cos\left[\left(\frac{2\pi a}{\lambda f}\right)\alpha_4 v\right]\right\}$$

$$-\left[\frac{\sin(\pi\alpha_5 bu)}{\pi\alpha_5 bu}\right]\left[\frac{\sin(\pi\beta_5 av)}{\pi\beta_5 av}\right] \times \left\{1 + 2\left[\frac{\sin(\pi\alpha_5 bu)}{\pi\alpha_5 bu}\right]\left[\frac{\sin(\pi\beta_5 av)}{\pi\beta_5 av}\right]\cos\left[\left(\frac{2\pi}{\lambda f}\right)\alpha_5 v\right]\right\}$$

$$(2.101)$$

We rewrite Eq. (2.101) in a summation form as follows:

$$h(u, v)$$
$$= \left[\frac{\sin(\pi bu)}{\pi bu}\right]\left[\frac{\sin(\pi av)}{\pi av}\right] \times \left\{1 + 2\left[\frac{\sin(\pi bu)}{\pi bu}\right]\left[\frac{\sin(\pi av)}{\pi av}\right]\cos\left[\left(\frac{4\pi a}{3\lambda f}\right)v\right]\right\}$$

$$+ \sum_{i=1}^{N-1}(-1)^i\left[\frac{\sin(\pi\alpha_i bu)}{\pi\alpha_i bu}\right]\left[\frac{\sin(\pi\beta_i av)}{\pi\beta_i av}\right] \times \{1$$

$$+2\left[\frac{\sin(\pi\alpha_i bu)}{\pi\alpha_i bu}\right]\left[\frac{\sin(\pi\beta_i av)}{\pi\beta_i av}\right]\cos\left[\left(\frac{2\pi}{\lambda f}\right)\alpha_i v\right]\} \qquad (2.102)$$

Where $i = 1, 2, 3, \ldots, N - 1$ for $N = 6$.

We fabricated a B/W hexagonal pupil where the number of zones $N = 6$ and the corresponding line plot at $y = 256$ pixels is shown in Fig. 2.22. The aperture is resized in a 512×512-pixel matrix, and the outer side length is $a = 126$ pixels, while the internal black hexagon has a side length equal to $a/6 = 21$ pixels.

Figure 2.23 shows the black and white concentric hexagonal apertures of the six zones and the corresponding normalized PSF. The total bandwidth of the central peak $= 4$ pixels. We computed the PSF using the fast Fourier transform (FFT). We computed the PSF for the uniform hexagonal aperture for comparison, as shown in Fig. 2.24.

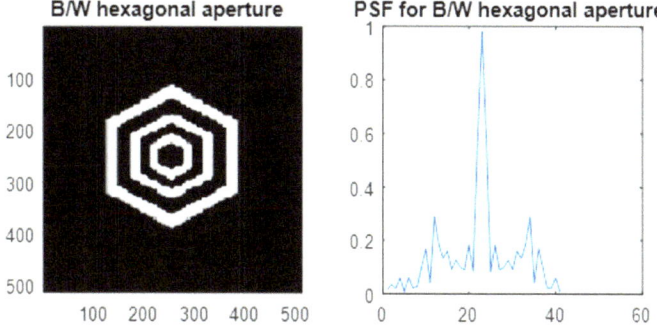

Fig. 2.23 A black and white concentric hexagonal aperture of six zones and the corresponding normalized PSF. The total bandwidth of the central peak = 4 pixels

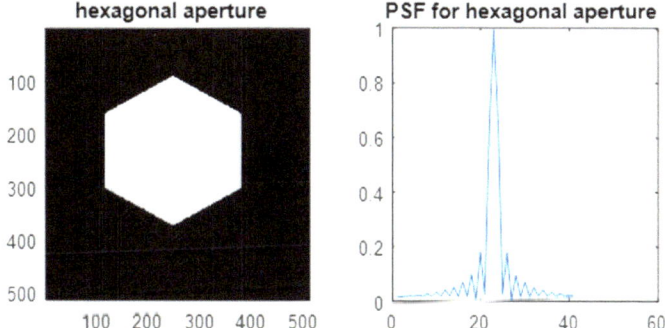

Fig. 2.24 A uniform hexagonal aperture and the corresponding normalized PSF. The total width of the central peak = 4 pixels. The FWHM is invariant for uniform hexagonal and modulated hexagonal apertures considering that the apertures are the same width

2.13 Four Hamming and Four Quadratic Apertures Around the Center Surrounded by an Annulus

The PSF corresponding to the aperture composed of a four Hamming and four quadratic apertures around the center surrounded by an annulus is computed as follows:

we assume that four circles of the Hamming function are arranged along the Cartesian coordinates, while another four circles rotated with 45° to the Cartesian coordinates varied quadratically. The eight circles around the center are surrounded by an annulus of a width $\delta\rho$ as shown in Fig. 2.25a. This aperture P_{H-Q}, is represented as follows since all the segments of the aperture are completely separated and located in the plane (u, v):

$$P_{H-Q}(\rho) = P_H(\rho) + P_Q(\rho) + P_{\text{annul}}(\rho) \tag{2.103}$$

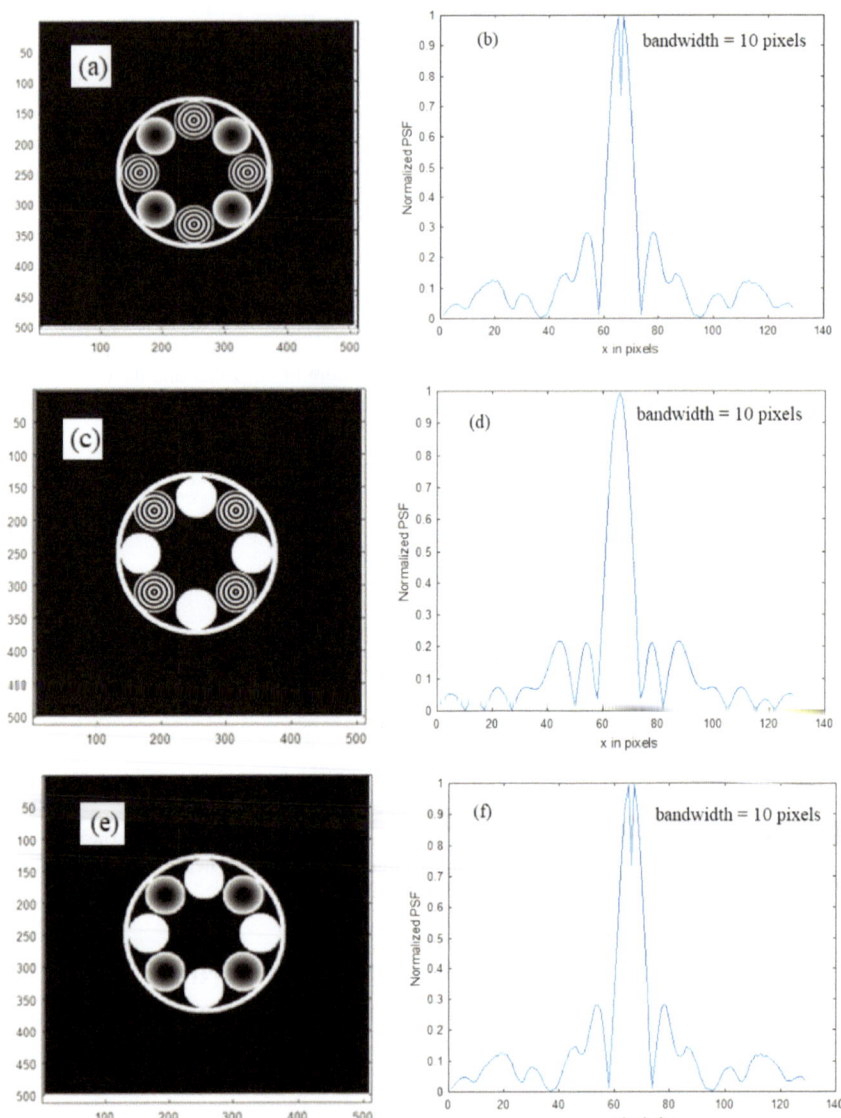

Fig. 2.25 The first three models' configurations and the corresponding normalized PSF plot, in the range [960–1088] in pixels, are shown. The external radius of all apertures = 128 pixels, and the radius of each of the eight small circles = 32 pixels. The annular width = 8 pixels. The sample size in PSF plots = 64 pixels for the total number of samples = 2048 pixels

The radial coordinate in the aperture plane is $\rho = \sqrt{u^2 + v^2}$.

The four Hamming apertures are located along the Cartesian coordinates at distances $(u \pm u_0 \pm \delta u, v)$ and $(u, v \pm v_0 \pm \delta v)$, and are represented as follows:

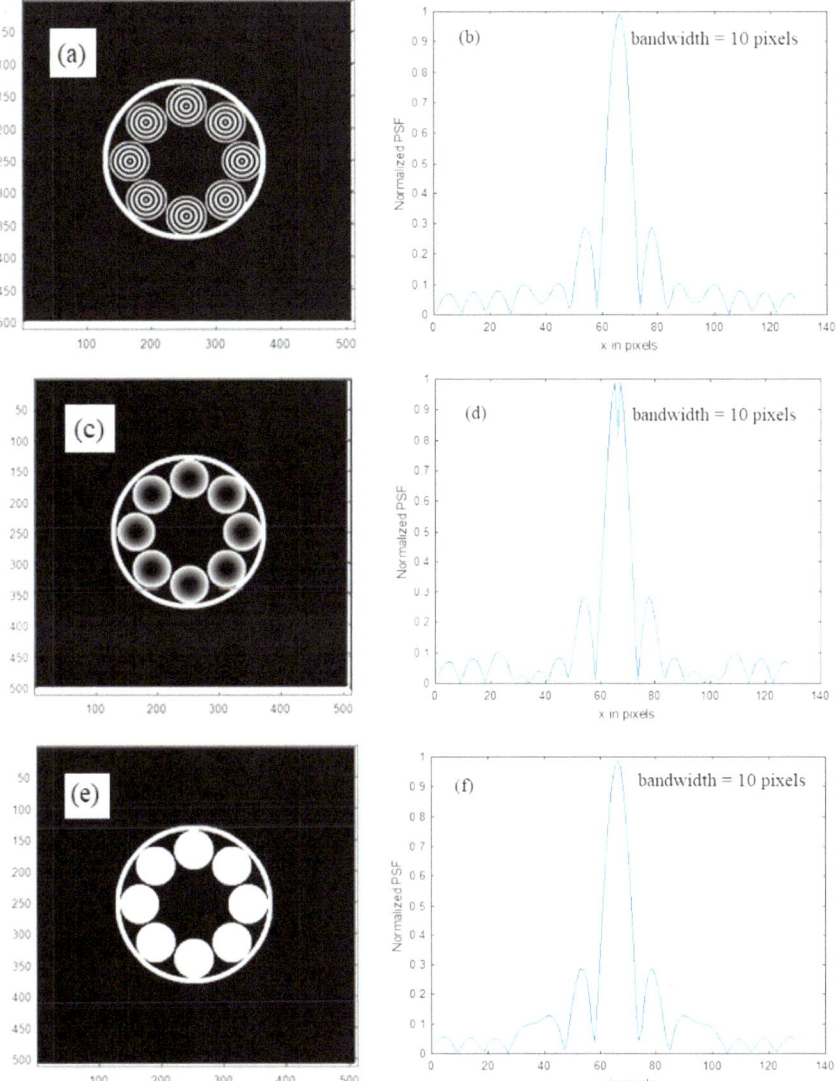

Fig. 2.26 Three aperture configurations and the corresponding normalized PSF plot, in the range [960–1088] in pixels are shown. The external radius of all apertures = 128 pixels, and the radius of each of the eight small circles = 32 pixels. The annular width = 8 pixels. The sample size in PSF plots = 64 pixels for the total number of samples = 2048 pixels

$$P_{\text{ham}}(\rho) = P_{H1}(\rho) + P_{H2}(\rho) + P_{H3}(\rho) + P_{H4}(\rho) \tag{2.104}$$

$$P_{H1}(\rho) = 0.54 + 0.46 * \cos\left(\sqrt{(u - u_0 - \delta u)^2 + v^2}\right) \tag{2.105}$$

$$P_{H2}(\rho) = 0.54 + 0.46 * \cos\left(\sqrt{(u + u_0 + \delta u)^2 + v^2}\right) \tag{2.106}$$

$$P_{H3}(\rho) = 0.54 + 0.46 * \cos\left(\sqrt{u^2 + (v - v_0 - \delta v)^2}\right) \tag{2.107}$$

$$P_{H4}(\rho) = 0.54 + 0.46 * \cos\left(\sqrt{u^2 + (v + v_0 + \delta v)^2}\right) \tag{2.108}$$

The parameter β is set equal to 0.6 in Eqs. (2.105–2.108).

The four apertures of quadratic distributions are located along the rotated coordinates by angle of $45°$ at distances $\left(u - \frac{u_0}{\sqrt{2}} - \delta u, v - \frac{v_0}{\sqrt{2}} - \delta v\right)$, $\left(u + \frac{u_0}{\sqrt{2}} + \delta u, v + \frac{v_0}{\sqrt{2}} - \delta v\right)$, $\left(u - \frac{u_0}{\sqrt{2}} - \delta u, v + \frac{v_0}{\sqrt{2}} + \delta v\right)$ and $\left(u + \frac{u_0}{\sqrt{2}} + \delta u, v - \frac{v_0}{\sqrt{2}} - \delta v\right)$, where, $\delta u = \delta v = \delta \rho$ and are represented as follows:

$$P_{\text{Quad}}(\rho) = P_{Q1}(\rho) + P_{Q2}(\rho) + P_{Q3}(\rho) + P_{Q4}(\rho) \tag{2.109}$$

$$P_{Q1}(\rho) = \left(u - \frac{u_0}{\sqrt{2}} - \delta u\right)^2 + \left(v - \frac{v_0}{\sqrt{2}} - \delta v\right)^2 \tag{2.110}$$

$$P_{Q2}(\rho) = \left(u + \frac{u_0}{\sqrt{2}} + \delta u\right)^2 + \left(v + \frac{v_0}{\sqrt{2}} + \delta v\right)^2 \tag{2.111}$$

$$P_{Q3}(\rho) = \left(u - \frac{u_0}{\sqrt{2}} - \delta u\right)^2 + \left(v + \frac{v_0}{\sqrt{2}} + \delta v\right)^2 \tag{2.112}$$

$$P_{Q4}(\rho) = \left(u + \frac{u_0}{\sqrt{2}} + \delta u\right)^2 + \left(v - \frac{v_0}{\sqrt{2}} - \delta v\right)^2 \tag{2.113}$$

$$P_{\text{annular}}(\rho) = 1; \ \rho_0 - \delta\rho < \rho < \rho_0 \tag{2.114}$$

The annulus is the difference between two circles of radii ρ_0 and $\rho_1 = \rho_0 - \delta\rho$. The PSF corresponding to the annular aperture is equal to the following:

$$h_{\text{annul}}(w) = \frac{2J_1(w)}{w} - \varepsilon^2 \frac{2J_1(w_1)}{w_1}; \ \varepsilon = \frac{\rho_1}{\rho_0} = 1 - \frac{\delta\rho}{\rho_0} \tag{2.115}$$

where $J_1(w)$ is the Bessel function of the 1st order for the reduced coordinate variable w.

The reduced coordinates, corresponding to the external and internal circles (w, and w_1), are computed as follows:

$$w = \frac{2\pi}{\lambda f}\rho_0 r, \text{ and } w_1 = \frac{2\pi}{\lambda f}\varepsilon\rho_0 r \qquad (2.116)$$

$r\,(x,y)$ is the radial coordinate in the Fourier plane corresponding to the aperture of radius ρ_0, and ε is defined in equation (2.115). λ is the wavelength of the illumination and f being the focal length corresponding to the Fourier transform lens.

We compute PSF corresponding to the shifted circles of quadratic distributions as follows:

$$h_{Q1}(r) = \text{F.T.}\left\{\left(u - \frac{u_0}{\sqrt{2}} - \delta u\right)^2 + \left(v - \frac{v_0}{\sqrt{2}} - \delta v\right)^2\right\}$$

$$= \text{F.T.}\left\{(u^2 + v^2) \otimes \delta\left(u - \frac{u_0}{\sqrt{2}} - \delta u, v - \frac{v_0}{\sqrt{2}} - \delta v\right)\right\} \qquad (2.117)$$

The symbolic convolution product of two functions is written in integral form as follows:

$$f(u,v) \otimes g(u,v) = \int\int_{-\infty}^{\infty} f(u',v')g(u-u', v-v')du'dv' \qquad (2.118)$$

Since the F.T. of the convolution product is equal to the simple product of the Fourier transform corresponding to each term, then:

$$h_{Q1}(r) = \text{F.T.}(u^2 + v^2).\text{F.T.}\left\{\delta\left(u - \frac{u_0}{\sqrt{2}} - u, v - \frac{v_0}{\sqrt{2}} - v\right)\right\} \qquad (2.119)$$

$$\text{F.T.}(\rho^2) = 2\left\{\frac{J_1\left(\frac{w}{4}\right)}{\frac{w}{4}} - 2\frac{J_2\left(\frac{w}{4}\right)}{\left(\frac{w}{4}\right)^2}\right\}; \text{ ref.}[5, 17] \qquad (2.120)$$

$$\text{F.T.}\left\{\delta\left(u - \frac{u_0}{\sqrt{2}} - \delta u, v - \frac{v_0}{\sqrt{2}} - \delta v\right)\right\} = \exp\left\{-\frac{j2\pi}{\lambda f}\left[\left(\frac{u_0}{\sqrt{2}} + \delta u\right)x + \left(\frac{v_0}{\sqrt{2}} + \delta v\right)y\right]\right\} \qquad (2.121)$$

By substituting Equations (2.120) and (2.121) in Eq. (2.119) we obtain:

$$h_{Q1}(r) = 2\exp\left\{-\frac{j2\pi}{\lambda f}\left[\left(\frac{u_0}{\sqrt{2}} + \delta u\right)x + \left(\frac{v_0}{\sqrt{2}} + \delta v\right)y\right]\right\}.\left\{\frac{J_1(w')}{w'} - 2\frac{J_2(w')}{w'^2}\right\} \qquad (2.122)$$

$$\text{where } w' = \frac{w}{4}.$$

Similar expressions, for the PSF corresponding to the other quadratic apertures, are obtained as follows:

$$h_{Q2}(r) = 2 \exp\left\{ +\frac{j2\pi}{\lambda f}\left[\left(\frac{u_0}{\sqrt{2}} + \delta u \right)x + \left(\frac{v_0}{\sqrt{2}} + \delta v \right)y \right] \right\} \cdot \left\{ \frac{J_1(w')}{w'} - 2\frac{J_2(w')}{w'^2} \right\}$$
(2.123)

$$h_{Q3}(r) = 2 \exp\left\{ -\frac{j2\pi}{\lambda f}\left[\left(\frac{u_0}{\sqrt{2}} + \delta u \right)x - \left(\frac{v_0}{\sqrt{2}} + \delta v \right)y \right] \right\} \cdot \left\{ \frac{J_1(w')}{w'} - 2\frac{J_2(w')}{w'^2} \right\}$$
(2.124)

$$h_{Q4}(r) = 2 \exp\left\{ +\frac{j2\pi}{\lambda f}\left[\left(\frac{u_0}{\sqrt{2}} + \delta u \right)x - \left(\frac{v_0}{\sqrt{2}} + \delta v \right)y \right] \right\} \cdot \left\{ \frac{J_1(w')}{w'} - 2\frac{J_2(w')}{w'^2} \right\}$$
(2.125)

Hence, the PSF corresponding to the four quadratic apertures is written as follows:

$$h_{Quad}(r) = 4\left\{ \frac{J_1(w')}{w'} - 2\frac{J_2(w')}{w'^2} \right\}\left[\cos\left\{ \frac{2\pi}{\lambda f}\left[\left(\frac{u_0}{\sqrt{2}} + \delta u \right)x + \left(\frac{v_0}{\sqrt{2}} + \delta v \right)y \right] \right\} \right.$$
$$\left. + \cos\left\{ \frac{2\pi}{\lambda f}\left[\left(\frac{u_0}{\sqrt{2}} + \delta u \right)x - \left(\frac{v_0}{\sqrt{2}} + \delta v \right)y \right] \right\} \right]$$
(2.126)

Equation (2.126) is rewritten as follows:

$$h_{Quad}(r) = 8\left\{ \frac{J_1(w')}{w'} - 2\frac{J_2(w')}{w'^2} \right\} \cdot \cos\left\{ \frac{\pi}{\lambda f}\left[\left(\frac{u_0}{\sqrt{2}} + \delta u \right)\left(\frac{r}{\sqrt{2}} \right) \right] \right\}$$
$$\cos\left\{ \frac{\pi}{\lambda f}\left[\left(\frac{v_0}{\sqrt{2}} + \delta v \right)\left(\frac{r}{\sqrt{2}} \right) \right] \right\}$$
(2.127)

In Eq. (2.127), we replaced the Cartesian coordinate (x, y) in the Fourier plane with the corresponding radial coordinates, where $x = y = \frac{r}{\sqrt{2}}$ For circular symmetry of revolution.

We substitute the reduced coordinates in Eq. (2.116), and we put $\delta u = \delta v = \delta \rho$, and $u_0 = v_0 = \frac{\rho_0}{\sqrt{2}}$ To obtain the following:

$$h_{Quad}(w) = 8\left\{ \frac{J_1(w')}{w'} - 2\frac{J_2(w')}{w'^2} \right\} \cdot \cos^2\left\{ \frac{w}{4\sqrt{2}}\left[\left(1 + \frac{2\delta\rho}{\rho_0} \right) \right] \right\}$$
(2.128)

The PSF corresponding to the four apertures of Hamming distribution are computed from Eqs. (2.105–2.108), and we obtain the following [8]:

$$h_{\text{ham}}(r) = 2\left\{ \delta(r) + \frac{1}{2}\left[\delta\left(r - \frac{f}{2}\right) + \delta\left(r + \frac{f}{2}\right) \right] \right\}$$

$$\left[\cos\left\{ \frac{2}{f}[u_0 + u]x \right\} + \cos\left\{ \frac{2}{f}[v_0 + v]y \right\} \right] \tag{2.129}$$

Consequently, for the 1st model of apertures, the PSF corresponding to the whole aperture is computed by adding Eqs. (2.115, 2.128, and 2.129) to obtain the PSF as follows:

$$h(w) = \left[\frac{2J_1(w)}{w} - 2\frac{2J_1(w_1)}{w_1} \right] + 8\left\{ \frac{J_1(w')}{w'} - 2\frac{J_2(w')}{w'^2} \right\}$$

$$\cdot \cos^2\left\{ \frac{w}{4\sqrt{2}}\left[\left(1 + \frac{2}{\rho_0} \right) \right] \right\} + 4\{\delta(r)$$

$$+ \frac{1}{2}\left[\left(r - \frac{f}{2} \right) + \left(r - \frac{f}{2} \right) \right] \cos\left\{ \frac{w}{2}\left[\left(1 + \frac{\sqrt{2}}{\rho_0} \right) \right] \right\} \tag{2.130}$$

The design of an aperture composed of an annulus of width = 8 pixels, surrounded by four circles of quadratic distribution and four circles of Hamming distribution around the center, each of radius = 32 pixels is shown in Fig. 2.25a, where the external radius of the circle = 128 pixels. The PSF is computed using the FFT. Normalized PSF, in the range [960–1088] in pixels corresponding to the aperture in Fig. 2.25a, is plotted as in Fig. 2.25b. The design of another aperture composed of an annulus of width = 8 pixels, surrounded by four circles of Hamming distribution and other four transparent circles around the center, each of radius = 32 pixels is shown in Fig. 2.25c. The corresponding normalized PSF, in the range [960–1088] in pixels, is plotted in Fig. 2.26d. In addition, the design of an aperture composed of an annulus of width = 8 pixels, surrounded by four circles of quadratic distribution and other four transparent circles around the center, each of radius = 32 pixels is shown in Fig. 2.25e. The corresponding normalized PSF, in the range [960–1088] in pixels, is plotted in Fig. 2.25f. Other apertures of eight Hamming functions around the center as in Fig. 2.26a, eight quadratic distributions as in Fig. 2.26c, and the corresponding PSF plots are shown in Fig. 2.26b and Fig. 2.26d. The comparison of the five models with the eight transparent circles PSF results is shown in Fig. 2.26e, f. In addition, comparisons with the annular and circular apertures are shown in Fig. 2.27a–d. Referring to the PSF plots, we showed that the central lobe has a constant bandwidth = 10 pixels for all the apertures like the circular aperture, while it has bandwidth = 7.5 pixels in the case of the annular aperture shown in Fig. 2.27b. Hence, the resolution is the same for all the model's configurations like the resolution corresponding to the transparent circular aperture. The contrast in the case of the aperture models is much better than that corresponding to the annular aperture. Hence, the contrast is improved for all the models and the circular aperture. In addition, we showed that the legs of the PSF plots are all different depending on the shape of the aperture configurations. In the case of the models, we showed that the secondary peaks are

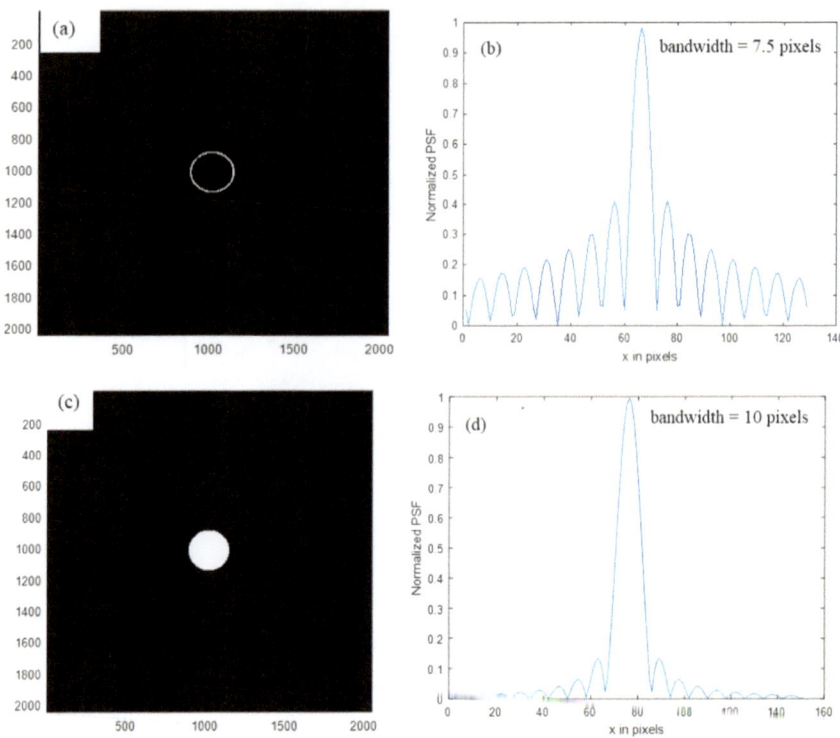

Fig. 2.27 The circular and annular apertures and the corresponding normalized PSF plots, in the central range [960–1088] in pixels. The external radius = 128 pixels, and the annular width = 8 pixels. The sample size in PSF plots = 64 pixels for the total number of samples = 2048 pixels

stronger compared to the circular PSF secondary peaks allowing us to image extended objects in microscopy.

The area is dependent on the aperture distribution at the defined range [960–1088]. For a transparent model configuration of 8 circles around the center surrounded by an annulus, we compute the area as follows:

$$A = \pi \left(r_0^2 - r_1^2 \right) + 8\pi \left(\frac{r_0}{8} \right)^2 = \pi \delta r (r_0 + r_1) + \frac{1}{8} \pi r_0^2$$

In the simulation, we assumed the annular width of the aperture $\delta r = 8$ pixels, $r_0 = 128$ pixles, and $r_1 = r_0 - \delta r = 120$ pixels. Hence, the total area = 38,403 and the area in the defined range = 12,666 × (1088 − 960)/2048 = 792 pixels.

The numerical models for the point spread function (PSF) of different aperture configurations are plotted using FFT as in Figs. 2.25, 2.26, and 2.27. The corresponding theoretical PSF curves are plotted in Fig. 2.28a–c compared to the PSF curves in the case of circular and annular apertures Fig. 2.28d, e).

The ratio $\frac{\delta \rho}{\rho_0} = 0.0625$ and $\varepsilon = 0.9375$ for the annulus.

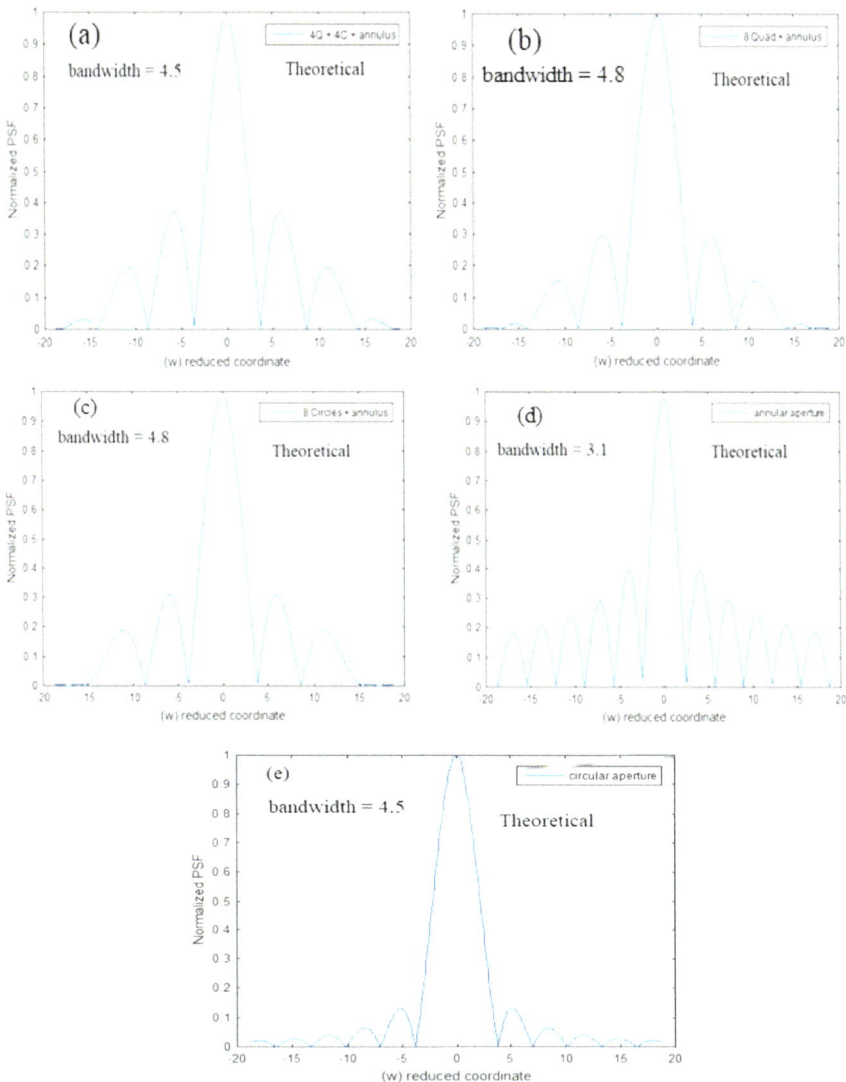

Fig. 2.28 Theoretical PSF plots computed from the analysis. The bandwidth is dimensionless computed from the full width at half maximum. The reduced coordinate w is dimensionless. In the computations, we set the annular width = 8 pixels compared to the external radius = 128 pixels

2.14 Conclusion

We showed that the PSF is dependent on the aperture distribution for the different models. We have assumed two different distributions inside the aperture arranged symmetrically as shown in Figs. 2.25, 2.26, and 2.27. In addition, the PSF legs or secondary peaks for the investigated models have stronger peaks than the corresponding PSF secondary weak peaks obtained in the case of transparent circular aperture. Also, the secondary peaks are different for the investigated apertures. Hence, these investigated apertures are useful for imaging extended objects. We obtained a constant bandwidth in the case of models' configurations like the circular aperture. Contrarily, we expected poorer contrast in the case of annular aperture compared to the improved contrast corresponding to the configuration models. Consequently, compromised resolution and contrast are attained for the presented models. We enhance the pixel resolution in the PSF plots taking a matrix of dimensions 2048 × 2048 pixels and aperture radius = 128 pixels, hence the sample size is increased to 64 pixels.

References

1. J.W. Goodman, *Introduction to Fourier Optics* (McGraw-Hill Book Comp, New York, 1968)
2. I.J. Cox, C.J.R. Sheppard, T. Wilson, Improvement in resolution by confocal microscopy. Appl. Opt. **21**, 778–781 (1982)
3. J.J. Clair, A.M. Hamed, Theoretical studies on optical coherent microscope. Optik **64**, 133–141 (1983)
4. A.M. Hamed, J.J. Clair, Image and super-resolution in optical coherent microscopes. Optik **64**, 272–284 (1983)
5. A.M. Hamed, J.J. Clair, Studies on optical properties of confocal scanning optical microscope using pupils with radially transmission distribution. Optik **65**, 209–218 (1983)
6. M. Castello, C.J.R. Sheppard et al., Image scanning microscopy with a quadrant detector. Opt. Lett. **40**(22), 5355–5358 (2015)
7. A.M. Hamed, Improvement of point spread function (PSF) using linear-quadratic aperture. Optik **131**, 838–849 (2017)
8. A.M. Hamed, T. Al-Saeed, Image analysis of modified Hamming aperture: application on confocal microscopy and holography. J. Modern Opt. **62**, 801–810 (2015)
9. A.M. Hamed, The point spread function of an aperture in the form of Corona Virus (COVID-19) images. Int. J. Emerging Eng. Res. Technol. (IJEERT) **8**(2), 17–22 (2020)
10. A.M. Hamed, Compromising of resolution and contrast in obstructed Cauchy aperture and its application in confocal microscopy. Opt. Quant. Electron. **55**, 1278 (2023)
11. A.M. Hamed, Design of a Cascaded Black-Linear Distribution (CBLD) in circular aperture and its application on Confocal Laser Scanning Microscope (CSLM). Am. J. Optics Photonics **7**(3), 46–56 (2019). https://doi.org/10.11648/j.ajop.20190703.11
12. A.M. Hamed, Investigation of a new modulated aperture using speckle techniques. J Basic Appl Sci **11**, 39 (2022). https://doi.org/10.1186/s43088-022-00222-2
13. A.M. Hamed, Discrimination between speckle images using diffusers modulated by some deformed apertures: simulations. Opt. Eng. **50**, 1–7 (2011)
14. A.M. Hamed, T. Al-Saeed, Investigation of rectangular apertures and their application on speckle imaging. J. Basic Appl. Sci. **11**(42) (2022)
15. A.M. Hamed, Application of a hexagonal aperture on the confocal scanning laser microscope. Opt. Quant. Electron. **55**, 749 (2023)
16. A.M. Hamed, Investigation of a concentric black and white hexagonal pupil and its application in confocal microscopic imaging. Opt. Quant. Electron. **55**, 1279 (2023)

Chapter 3
Coherent Transfer Functions (CTFs) for Some Modulated Apertures

3.1 Computation of CTF for Some Modulated Apertures

The coherent transfer function CTF is the convolution product of the apertures P_2 and P_2 and is represented in a symbolic form as follows:

$$\text{CTF}(u, v) = P_1(u, v) \otimes P_2(u, v) \tag{3.1}$$

$\otimes$: is a symbol for convolution.

For two equal objectives, the CTF is replaced by autocorrelation. We plot the CTF corresponding to some modulated apertures.

We computed the CTF for the CSLM from the autocorrelation of the pupil function.

3.1.1 CTF for the Modulated Cauchy Aperture [1, 2]

The Cauchy aperture is mathematically represented as follows:

$$P_{\text{Cauchy}}(\rho) = \frac{1}{1 + \beta^2 \left(\frac{\rho}{\rho_0}\right)^2}; \left|\frac{\rho}{\rho_0}\right| \leq 1, \text{ and } \beta < 1 \tag{3.2}$$

β is a parameter, and ρ is the radial coordinate in the aperture plane with a maximum radius of ρ_0 normalized equal unity.

We write the CTF corresponding to the Cauchy aperture in integral form as follows:

A. M. Hamed, *Studies on the Confocal Laser Microscope*,
SpringerBriefs in Applied Sciences and Technology,
https://doi.org/10.1007/978-3-031-87275-4_3

$$\text{CTF}(\rho) = \int_{-\infty}^{\infty} \left[\frac{1}{1 + \beta^2 \rho'^2} \right] \left[\frac{1}{1 + \beta^2 (\rho - \rho')^2} \right] d\rho' \qquad (3.3)$$

We write the CTF symbolically as follows:

$$\text{CTF}(\rho) = \frac{1}{1 + (\beta \cdot \rho)^2} \otimes \frac{1}{1 + (\beta \cdot \rho)^2} \qquad (3.4)$$

We represent the obstructed Cauchy aperture as the difference between the ordinary Cauchy aperture represented in Eq. (3.2) and a circular central zone of maximum radius $= \frac{\rho_0}{4}$ as follows:

$$P_{\text{obst.}}(\rho) = P_{\text{cauchy}}(\rho) - \text{circ}\left(\frac{4\rho}{\rho_0} \right) \qquad (3.5)$$

Hence, we write the CTF, corresponding to the modulated Cauchy aperture Eq. (3.5), symbolically as follows:

$$\text{CTF}(\rho) = \left\{ \frac{1}{1 + (\beta \cdot \rho)^2} - \text{circ}\left(\frac{4\rho}{\rho_0} \right) \right\} \otimes \left\{ \frac{1}{1 + (\beta \cdot \rho)^2} - \text{circ}\left(\frac{4\rho}{\rho_0} \right) \right\} \quad (3.6)$$

3.1.2 CTF for Modulated Aperture (Eight Equally Spaced Conic Apertures) [3–5]

The equally spaced eight conic apertures shown in Fig. 3.2 are described as follows (Fig. 3.1):

$$P_T(u, v) = P_1 + P_2 + P_3 + P_4 + P_5 + P_6 + P_7 + P_8 \qquad (3.7)$$

With the apertures $P_{1,2}$ located along the x-axis at distances $= \pm u_0$, and the apertures $P_{3,4}$ located along the y-axis at distances $= \pm v_0$ represented as follows:

$$P_{1,2}(u, v) = 1 - \sqrt{(u \pm u_0)^2 + v^2} \qquad (3.8)$$

$$P_{3,4}(u, v) = 1 - \sqrt{u^2 + (v \pm v_0)^2} \qquad (3.9)$$

For rotated coordinates by an angle of 45°, another four apertures are shown in the same Fig. Fig. 3.3 and represented as follows:

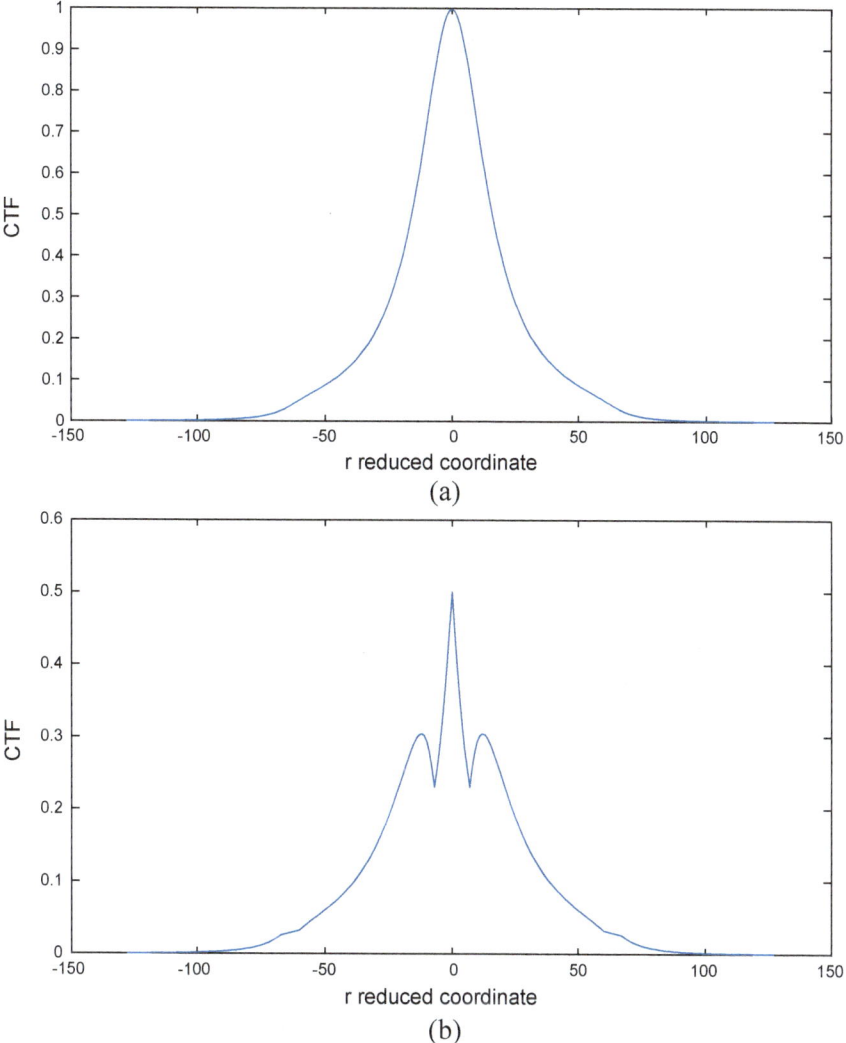

Fig. 3.1 a The plot of the normalized CTF or the autocorrelation corresponding to the Cauchy aperture represented in Eq. (3.2). **b** The plot of the normalized CTF corresponding to the obstructed Cauchy aperture. The aperture radius $= 64$ pixels, and the obstruction central zone radius $= 4$ pixels

$$P_{5,6}(u, v) = 1 - \sqrt{\left(u \pm \frac{u_0}{\sqrt{2}}\right)^2 + \left(v \pm \frac{v_0}{\sqrt{2}}\right)^2} \tag{3.10}$$

$$P_7(u, v) = 1 - \sqrt{\left(u - \frac{u_0}{\sqrt{2}}\right)^2 + \left(v + \frac{v_0}{\sqrt{2}}\right)^2} \tag{3.11}$$

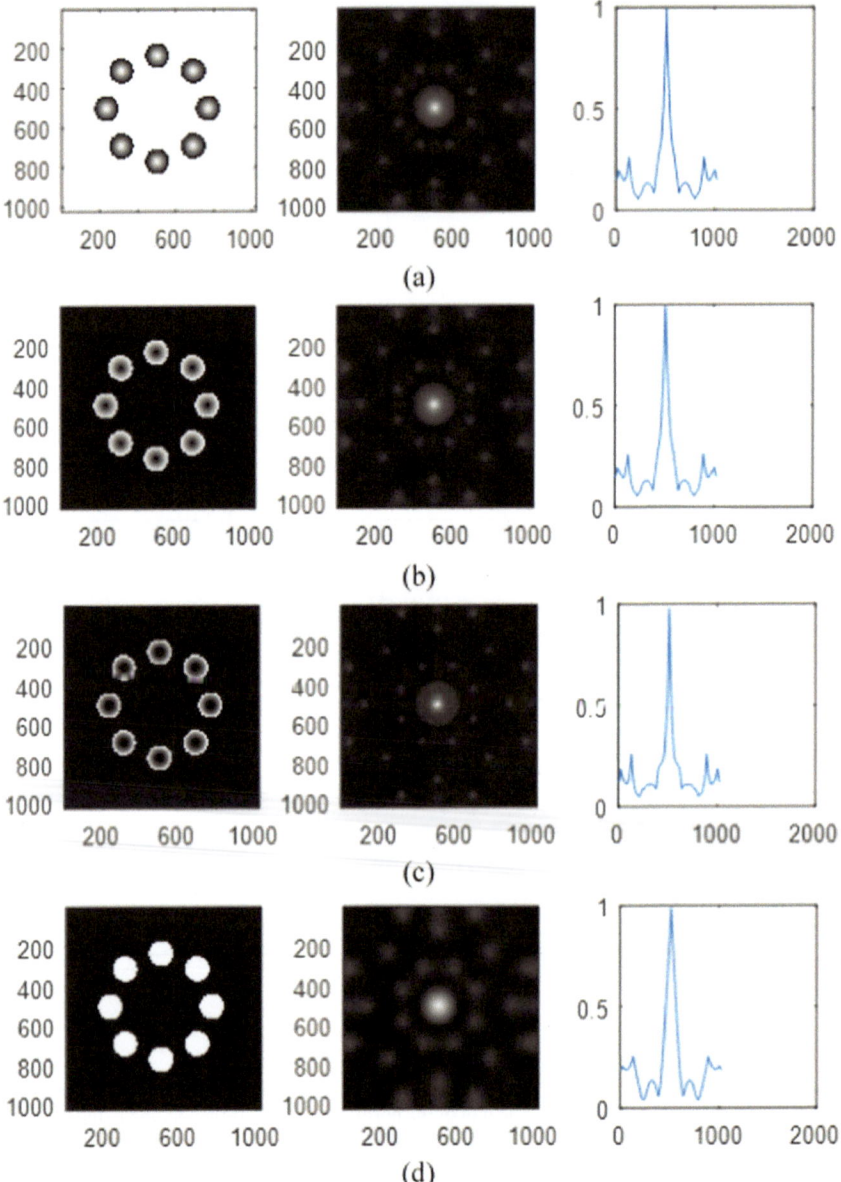

Fig. 3.2 a Autocorrelation image and plot of the conical aperture. **b** Autocorrelation image and plot of the linear aperture. **c** Autocorrelation image and plot of the quadratic aperture. **d** Autocorrelation image and plot for the uniform circular aperture

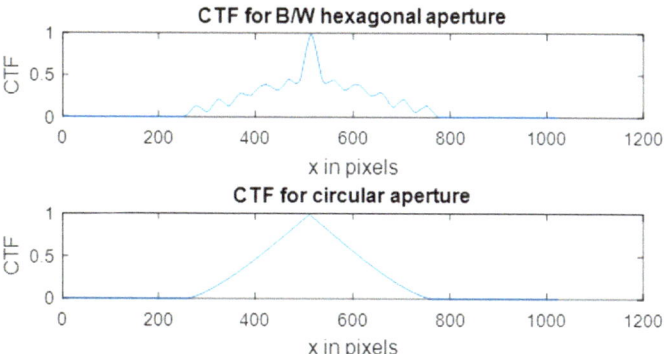

Fig. 3.3 We plotted the autocorrelation corresponding to the B/W hexagonal aperture compared with that corresponding to the uniform circular aperture. The autocorrelation bandwidth is two times the maximum diameter of the apertures [6, 7]

Fig. 3.4 For the Hamming aperture, the CTF is shown [8]. The aperture has a total width of 16 pixels, and the CTF has a total width of 32 pixels

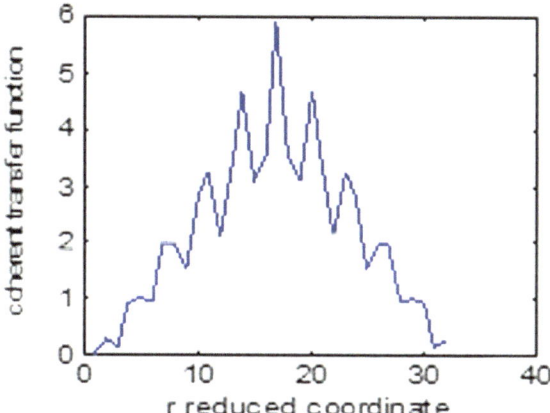

$$P_8(u, v) = 1 - \sqrt{\left(u + \frac{u_0}{\sqrt{2}}\right)^2 + \left(v - \frac{v_0}{\sqrt{2}}\right)^2} \qquad (3.12)$$

The CTF, corresponding to the aperture described by formulae (3.7–(3.12), is computed from Eq. (3.1) and is represented in Fig. 3.2.

3.2 Results and Discussion

Using the formula (3.3), we computed the CTF for the Cauchy aperture which is plotted in Fig. 3.1a. We computed the CTF, corresponding to the modulated Cauchy aperture from Eq. (3.6), and is plotted in Fig. 3.1b.

References

1. J.W. Goodman, *Introduction to Fourier Optics* (McGraw-Hill Book Comp, New York, 1968)
2. A.M. Hamed, Compromising of resolution and contrast in obstructed Cauchy aperture and its application in confocal microscopy. Opt. Quant. Electron. **55**, 1278 (2023)
3. A.M. Hamed, J.J. Clair, Image and super-resolution in optical coherent microscopes. Optik **64**, 272–284 (1983)
4. A.M. Hamed, J.J. Clair, Studies on optical properties of confocal scanning optical microscope using pupils with radially transmission distribution. Optik **65**, 209–218 (1983)
5. A.M. Hamed, Investigation of a new modulated aperture using speckle techniques. J Basic Appl Sci **11**, 39 (2022). https://doi.org/10.1186/s43088-022-00222-2
6. A.M. Hamed, Application of a hexagonal aperture on the confocal scanning laser microscope. Opt. Quant. Electron. **55**, 749 (2023)
7. A.M. Hamed, Investigation of a concentric black and white hexagonal pupil and its application in confocal microscopic imaging. Opt. Quant. Electron. **55**, 1279 (2023)
8. A.M. Hamed, T. Al-Saeed, Image analysis of modified Hamming aperture: application on confocal microscopy and holography. J. Modern Opt. **62**, 801–810 (2015)

Chapter 4
The Imaging of Microscopic Objects Using a Confocal Microscope

4.1 Introduction

In this chapter, we summarize the process of image formation using a confocal microscope that provides modulated apertures. We plotted the reconstructed images using Cauchy and hexagonal apertures under a confocal scanning laser microscope. Second, I suggested an optical correlator using a confocal microscope. We provide some correlation images corresponding to the ideal case of coherent illumination and coherent detection in a confocal microscope.

4.2 Imaging Using Modulated Apertures via Confocal Microscopy

The algorithm used to compute the image in the CSLM is summarized as follows [1–4]:

1. Compute the PSF for objectives L_1 and L_2 as follows:

$$h_1(x, y) = \text{F.T. } \{P_1(u, v)\} \text{ and } h_2 = \text{F.T.}\{P_2(u, v)\} \qquad (4.1)$$

2. Both PSFs are multiplied to obtain the resultant PSF as follows:

$$h_r(x, y) = h_1(x, y) \cdot h_1(x, y) \qquad (4.2)$$

3. The inverse of the resultant PSF is taken to obtain the coherent transfer function CTF as the convolution product as follows:

$$\text{CTF}(u, v) = P_1(u, v) \otimes P_2(u, v) \qquad (4.3)$$

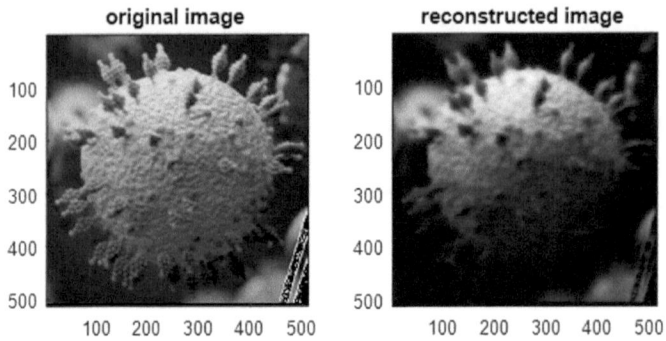

Fig. 4.1 Reconstructed image for the coronavirus image in the CSLM. The two objectives provided obstructed Cauchy apertures [1]. The obstruction central zone of radius $= 4$ pixels, and $\beta = 0.5$

4. Compute the Fourier spectrum of the object as follows:

$$G(u, v) = \text{F.T.}\{g(x, y)\} \tag{4.4}$$

5. The CTF and the Fourier spectrum of the object are multiplied to obtain the following:

$$\text{CTF}(u, v) \cdot G(u, v) \tag{4.5}$$

6. Take the inverse Fourier transform of (4.5) to obtain:

$$A(x, y) = \left[h_1(x, y) \times h_2(x, y) \right] \otimes g(x, y) \tag{4.6}$$

7. Finally, the modulus square of (4.6) is taken to obtain the intensity detected for the CSLM as:

$$I(x, y) = |h_1(x, y) h_2(x, y) \otimes g(x, y)|^2 \tag{4.7}$$

We plotted the reconstructed images using Cauchy [1] and hexagonal apertures [2] under a confocal scanning laser microscope (Figs. 4.1 and 4.2).

4.3   Image Processing Using a Confocal Scanning Laser Microscope

Referring to Fig. 4.3, transparency O_1 moves in the plane (x, y), while the second transparency O_2, which is found at a distance ϵ from the first one, is stationary. We assume that O_2 is much closer to O_1, which permits us to write: $O_2(x_0, y_0) = O_2(x, y)$. Hence, the intensity distribution in the imaging plane is calculated to give [5] the

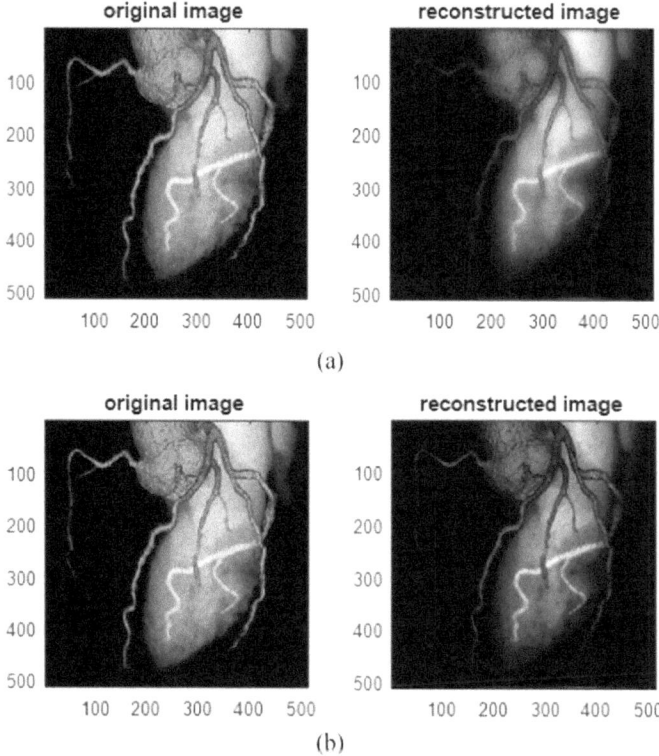

Fig. 4.2 a The reconstructed image of the coronary arteries using the CSLM [2] provided with the two objectives of the microscope with B/W hexagonal apertures of external sides = 128 pixels. **b** The reconstructed image of the coronary arteries using the CSLM [2] provided with the two objectives of the microscope with uniform hexagonal apertures of side length = 128 pixels

following:

$$I(x_s, y_s) = |O_1 \otimes h_1 h_2 O_2 \exp\left(\frac{jkd^2}{2 \in}\right)|^2(x_s, y_s) \tag{4.8}$$

where $d = \sqrt{x_0^2 + y_0^2}$ and $k = 2\pi/\lambda$.

Equation (4.8) can be expanded to yield:

$$I(x_s, y_s) = |O_1 \otimes h_1 h_2 O_2 \cos\left(\frac{\pi d^2}{\lambda \in}\right)|^2 + |O_1 \otimes h_1 h_2 O_2 \sin\left(\frac{\pi d^2}{\lambda \in}\right)|^2 \tag{4.9}$$

For perfect imaging, which is the case of geometrical optics approximation, we have the following: $h_1 h_2 = 1$, and in the case where $\frac{\pi d^2}{\lambda \in} \ll 1$, Eq. (4.9) gives:

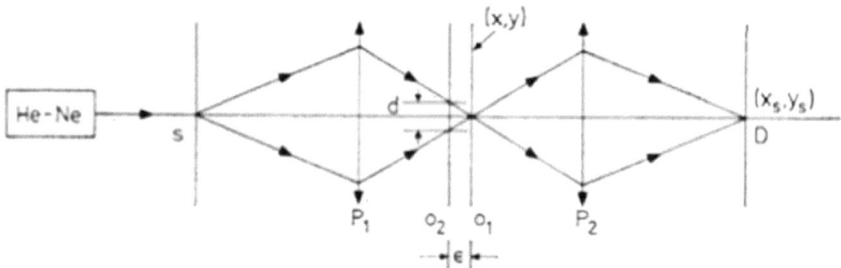

Fig. 4.3 Image processing (correlation of O_1, and O_2) using a confocal laser microscope

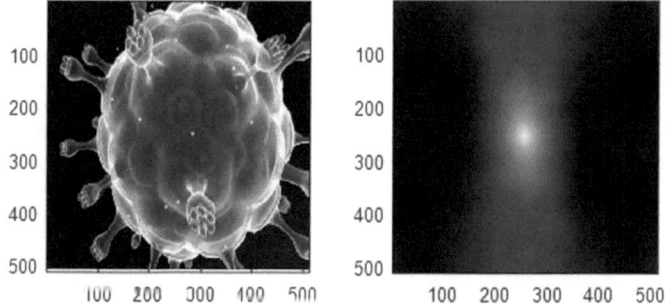

Fig. 4.4 The SIDA virus image and the corresponding autocorrelation image

$$I(x_s, y_s) = |O_1 \otimes O_2|^2 + \left| O_1 \otimes O_2 \cdot \left(\frac{\pi d^2}{\lambda \in} \right) \right|^2 \tag{4.10}$$

The first term in Eq. (4.10) gives the correlation intensity between the two signals O_1 and O_2, while the second term is considered noise. Hence, to improve the correlation results, it is preferable to place the two transparencies as close as possible in the plane (x, y). Consequently, the second term vanishes, and we obtain:

$$I(x_s, y_s) = |O_1 \otimes O_2|^2 \tag{4.11}$$

We plot some correlation images corresponding to coherent illumination and coherent detection in the microscope, hence obtaining the exact correlation intensity Eq. (4.11). We obtain the correlation intensity for some objects represented in Figs. 4.4, 4.5, 4.6, 4.7 and 4.8. All images have dimensions of 512×512 pixels.

Fig. 4.5   The image of blood cells (erythrocytes) and the corresponding autocorrelation image

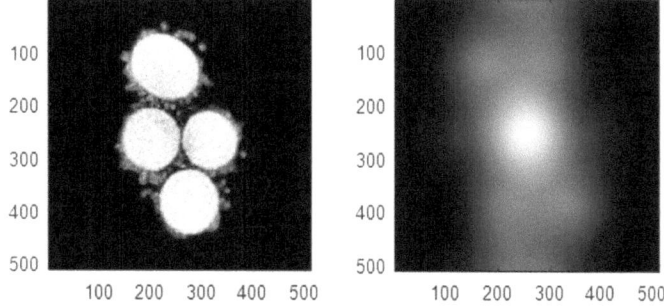

Fig. 4.6   COVID-19 image and the corresponding autocorrelation image

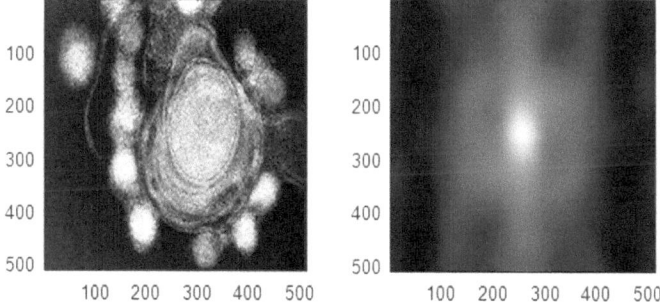

Fig. 4.7   The image of Zika brain cancer and the corresponding autocorrelation image

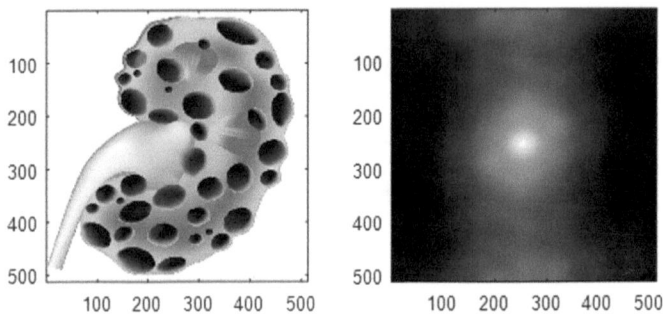

Fig. 4.8 Image of kidney damage and the corresponding autocorrelation image

References

1. A.M. Hamed, Compromising of resolution and contrast in obstructed Cauchy aperture and its application in confocal microscopy. Opt. Quant. Electron. **55**(14), 1278 (2023). https://doi.org/10.1007/s11082-023-05507-z
2. A.M. Hamed, Investigation of a concentric black and white hexagonal pupil and its application in confocal microscopic imaging. Opt. Quant. Electron. **55**, 1279 (2023)
3. A.M. Hamed, Design of a Cascaded Black-Linear Distribution (CBLD) in circular aperture and its application on Confocal Laser Scanning Microscope (CSLM). Am. J. Optics Photonics **7**(3), 46–56 (2019). https://doi.org/10.11648/j.ajop.20190703.11
4. A.M. Hamed, Modulated apertures and resolution in microscopy. Springer Briefs in Applied Sciences and Technology (2023). ISBN 978-3-031-47552-8 (eBook)
5. J.J. Clair, A.M. Hamed, Theoretical studies on optical coherent microscope. Optik **64**, 133–141 (1983)

Chapter 5
Theoretical Study on a Coherent Non-scanned Laser Microscope (CNSM)

5.1 Introduction

In this chapter, I present a confocal microscope that works without mechanical scanning of the object. I named it a confocal non-scanned laser microscope. I investigated it theoretically using two identical gratings instead of mechanical scanning [4], and the reconstruction of images in a non-scanned confocal microscope (NSCM) using speckle imaging is given in [5].

5.2 Theoretical Analysis

This microscope, namely, CNSM, is a coherent microscope that works without mechanical scanning of the object. Referring to Fig. 5.1. Instead of performing mechanical scanning of the object, we fix it in place and place two identical gratings $G_{1,2}$ very close to the objectives L_1 and L_2 to replace the scanning. In this case, the spreading of the Fourier spectrum of the grating functions upon the object will cover the whole object.

Assuming coherent light is emitted from the laser beam, the complex amplitude transmitted from the grating, composed of a periodic train of unit impulses, is mathematically represented as:

$$G_i(u, v) = \sum_{n=-\infty}^{\infty} \sum_{m=-\infty}^{\infty} \delta(u - nu_0, v - mv_0) \tag{5.1}$$

where $i = 1$ represents the 1st grating, $i = 2$ represents the 2nd grating, or in a more elaborate form,

© The Author(s), under exclusive license to Springer Nature Switzerland AG 2025
A. M. Hamed, *Studies on the Confocal Laser Microscope*,
SpringerBriefs in Applied Sciences and Technology,
https://doi.org/10.1007/978-3-031-87275-4_5

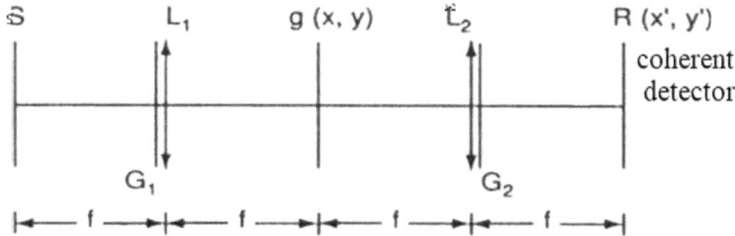

Fig. 5.1 Coherent Non-Scanned Laser Microscope (CNSLM). S: point source, D: point detector, g: transparent object situated in the plane (x, y), G_1, G_2: two identical gratings each composed of a periodic train of unit impulses

$$G_i(u, v) = \sum_{n=-\infty}^{\infty} \sum_{m=-\infty}^{\infty} \text{circ}(u - nu_0, v - mv_0) \qquad (5.2)$$

where $\rho_0 = \sqrt{u_0^2 + v_0^2}$ is the pinhole radius.

The Fourier transform of the grating comb function is another comb function, i.e.,

$$g_i(x, y) = \text{F.T.}[G_i(u, v)] = \sum_{n=-\infty}^{\infty} \sum_{m=-\infty}^{\infty} \text{circ}\left(x' - \frac{n}{u_0}, y' - \frac{m}{v_0}\right) \qquad (5.3)$$

The point spread function of a clear circular aperture is:

$$h_i(w) = 2\frac{J_i(w)}{w}; i = 1, 2 \qquad (5.4)$$

where $i = 1$ stands for the 1st pupil, and $i = 2$ for the 2nd pupil.

The PSF in the CSM is $h_r = h_1 h_2$, , while in the presence of the grating functions placed very close to the apertures, the effective point spread function (h_{eff}) is obtained by operating F.T. on the simple product of the pupil function and the grating function, hence, we obtain:

$$h_{1\,\text{eff}}(x, y) = \text{F.T.}[P_1(u, v) \cdot G_1(u, v)] = h_1(x, y) \otimes g_1(x, y) \qquad (5.5)$$

A similar expression is obtained for the 2nd aperture:

$$h_{2\,\text{eff}}(x, y) = \text{F.T.}[P_2(u, v) \cdot G_2(u, v)] = h_2(x, y) \otimes g_2(x, y) \qquad (5.6)$$

Operating the convolution product Eq. (5.3), $h_{1\,\text{eff}}$ becomes:

$$h_{1\,\text{eff}}(x, y) = \sum_{n=-\infty}^{\infty} \sum_{m=-\infty}^{\infty} \delta\left(x - \frac{n}{u_0}, y - \frac{m}{v_0}\right) \otimes h_1(x, y)$$

$$= \sum_{n=-\infty}^{\infty} \sum_{m=-\infty}^{\infty} h_1\left(x - \frac{n}{u_0}, y - \frac{m}{v_0}\right) \tag{5.7}$$

A similar expression is obtained for $h_{2\,\text{eff}}$ as follows:

$$h_{2\,\text{eff}}(x, y) = \sum_{n=-\infty}^{\infty} \sum_{m=-\infty}^{\infty} h_2\left(x - \frac{n}{u_0}, y - \frac{m}{v_0}\right) \tag{5.8}$$

To overcome the interference problem that occurs from the adjacent pinholes of the grating, this condition must be fulfilled:

$$r_d = \lambda f \rho_0 / u_0 v_0 \tag{5.9}$$

where r_d is the radial distance of the diffraction spot corresponding to each objective lens, and f is the focal length of L_1 and L_2.

The intensity distribution in the detector plane (x', y') of a CSM is given as [1–3]:

$$I(x', y', x_s, y_s) = \left| \int \int h_{1\,\text{eff}}(x, y) g(x - x_s, y - y_s) h_{2\,\text{eff}}(x + x', y + y') dx\, dy \right|^2 \tag{5.10}$$

where (x_s, y_s) is the scan position of the object.

When an axis point detector is employed $(x' = y' = 0)$, Eq. (5.10) becomes

$$I(x_s, y_s) = \left| \int \int h_{1\,\text{eff}}(x, y) h_{2\,\text{eff}}(x, y) g(x - x_s, y - y_s) dx dy \right|^2 \tag{5.11}$$

In the absence of scanning, Eq. (5.10) reduces this integral to

$$I(x_s = 0, y_s = 0) = \left| \int \int h_{1\,\text{eff}}(x, y) h_{2\,\text{eff}}(x, y) g(x, y) dx dy \right|^2 \tag{5.12}$$

Substitute Eqs. (5.7) and (5.8) in Eq. (5.12) we get:

$$I\left(\frac{n}{u_0}, \frac{m}{v_0}\right) = \left| \int \int \sum_{n=-\infty}^{\infty} \sum_{m=-\infty}^{\infty} h_r\left(x - \frac{n}{u_0}, y - \frac{m}{v_0}\right) g(x, y) dx dy \right|^2 \tag{5.13}$$

This can be written in compact form as follows [4]:

$$I\left(\frac{n}{u_0}, \frac{m}{v_0}\right) = \left| \sum_{n=-\infty}^{\infty} \sum_{m=-\infty}^{\infty} g\left(\frac{n}{u_0}, \frac{m}{v_0}\right) \otimes h_r\left(\frac{n}{u_0}, \frac{m}{v_0}\right) \right|^2 \tag{5.14}$$

Comparing Eq. (5.14) with that obtained by Sheppard via CSLM:

$$g(x, y) = \sum_{n=-\infty}^{\infty} \sum_{m=-\infty}^{\infty} g\left(\frac{n}{u_0}, \frac{m}{v_0}\right) \tag{5.15}$$

Hence, the scanning is realized in CNSLM through the manipulation of the sampling that occurs in the object plane. Consequently, we have achieved scanning of the object through manipulation of the periodic train of unit impulses placed in a plane very close to the objectives. The whole spatial information recorded in the imaging plane (x', y') is constructed using a coherent detector composed of an array matrix of two-dimensional point detectors. This matrix is represented as follows:

$$R(x', y') = \sum_{n=-\infty}^{\infty} \sum_{m=-\infty}^{\infty} \delta(x' - nx_0, y' - my_0) \tag{5.16}$$

where $x_0 = \frac{1}{u_0}$, $y_0 = \frac{1}{v_0}$

The manipulation of both gratings G_1 and G_2 placed nearer to the objectives L_1 and L_2 is recommended to realize the convolution operation of both objects and the resultant point spread function, while the manipulation of a single crossed grating placed behind the first lens L_1 does not produce the recommended convolution since h_2 is still unchanged while h_1 becomes:

$$h_{1\,eff} = h_1 \otimes g_1.$$

References

1. C.J.R. Sheppard, A. Choudhary, Opt. Acta **24**, 1051–1059 (1977)
2. I.J. Cox, C.J.R. Sheppard, T. Wilson, Improvement in resolution by confocal microscopy. Appl. Opt. **21**, 778–781 (1982)
3. J.J. Clair, A.M. Hamed, Theoretical studies on optical coherent microscope. Optik **64**, 133–141 (1983)
4. A.M. Hamed, Theoretical study on a Coherent Non-Scanned Microscope (CNSM). Optik **107**, 89–92 (1998)
5. A.M. Hamed, T. Al-Saeed, Reconstruction of images in a non-scanned confocal microscope (NSCM) using speckle imaging. J. Basic Appl. Sci. **10**, 67 (2021)

Chapter 6
Computation of the Lateral and Axial Point Spread Functions in Confocal Imaging Systems Using Binary Amplitude Mask

In this chapter, a novel aperture based on the TOLARDO concept composed of a central clear disk surrounded by a series of black and white (B/W) concentric annuli of equal transmittance is presented. Different apodized apertures of different numbers of B/W annuli are suggested to improve further the three-dimensional resolving power of confocal imaging systems. Both the axial and lateral point spread functions (PSF) and the corresponding irradiances are computed in both cases of conventional and confocal scanning microscopes for the above-mentioned amplitude filters. These results of axial and lateral irradiances are graphically represented by constructing a computer program using MATLAB. The obtained results are compared with those obtained in the case of circular, annular, and Martinez-Corral apodized aperture.

6.1 Introduction

The transverse resolution of the coherent scanning optical microscope (CSOM) is dependent on the apertures of the objective lenses. This microscope [1–5] is composed of two conjugate objective lenses arranged in tandem where the transparent object is located at the common short focus of the objectives. The mechanical scanning of the object is synchronized with the electronic scanning of the detector to construct the image during its scanning. It is usual to mention that the point source and the point detector are responsible for the spatial coherence of the illumination and the detection [6–10]. The theory of this confocal microscope showed that the resultant point spread function is calculated as the product of the point spread functions corresponding to both objective lenses. The main advantage of this confocal microscope is its ability to give better resolution when compared with conventional optical microscopes.

Another advantage of the confocal microscope is its ability to suppress the legs of its irradiance distribution. Further improvements in transverse resolution are attained

© The Author(s), under exclusive license to Springer Nature Switzerland AG 2025
A. M. Hamed, *Studies on the Confocal Laser Microscope*,
SpringerBriefs in Applied Sciences and Technology,
https://doi.org/10.1007/978-3-031-87275-4_6

[11–16] due to amplitude modulation designed on the apertures. Hence, most of the reported apodization techniques have been aimed at improving the resolving capability of conventional two-dimensional (2D) imaging systems, and therefore improving the transverse resolution [14, 15]. However, when dealing with confocal imaging systems, it is widely used because of its intrinsic optical sectioning capability of imaging three-dimensional (3D) objects. Recently, many authors have designed pupil filters based on the TOLARDO concept [17]. He showed that by subdividing the pupil into a proper number of concentric annular zones with constant transmittance, a band-limited transverse diffraction pattern of any shape may be obtained. His concept of modulation has been further applied to shaping the transverse PSF [16, 18, 19] and the axial PSF [20–24]. Consequently, both the transverse and axial resolutions are considered equally important for three-dimensional imaging.

Recently, different apodization techniques were proposed by many authors. Among these authors, are Cheng and Siu [25] and Siu et al. [26]. They used the apodization technique to achieve the suppression of the side lobes in the point spread function in the confocal scanning system. A combination of high numerical aperture of 1.45 oil immersion lens with super-resolving binary filters has been experimentally used to obtain an effective increase in axial resolution by measuring the full-width-half-maximum (FWHM). Also, a novel high-resolution scanning confocal microscope [25] with a high numerical aperture (NA = 1) parabolic mirror objective is investigated.

In this chapter, a novel pupil based on the TOLARDO concept is presented [27] A great number of black and white concentric equally spaced annuli of $N = 19$ with constant transmittance is considered to compute both the lateral and axial amplitude point spread function (APSF). Also, different arbitrarily selected manipulations of non-equally spaced annuli and hence unequal transmittance may be used to improve the resolution. The irradiance distributions of conventional and confocal microscopes are computed using our model of the multi-ring aperture of equally spaced annuli. Finally, the obtained results are compared with the corresponding results of the transverse and axial irradiances in the case of Martinez-Corral et al. [23] and the case of circular apertures. Also, no equally spaced annuli are investigated by changing the obscuration parameter μ. The obtained theoretical results of the APSF and the corresponding irradiances are plotted using MATLAB [28] and discussed, then followed by a conclusion.

6.2 Theoretical Analysis

Consider the amplitude point spread function (APSF) of a coherent imaging system apodized by a purely absorbing pupil filter. Hence [1], we write

$$h(u, w_{20}) = 2 \int_0^1 P(\rho) \exp\left(-j2\pi w_{20}\rho^2\right) J_0(2\pi u\rho)\rho d\rho, \qquad (6.1)$$

where $P(\rho)$ is the pupil function, ρ being the normalized radial coordinate in the pupil plane, $u = (\rho_0/\lambda f)r$ corresponds to the transverse radial coordinate expressed in optical units, ρ_0 being the maximum extent of the pupil, f is the focal length of the imaging lens and J_0 denotes the Bessel function of the 1st kind and zero order. $w_{20} = r_{20}/2\lambda f^2$ specifies the amount of defocus measured in units of wavelength and r_{20} is the actual radial distance of the defocused cross-section.

Next, it is convenient to perform the following non-linear mapping to separate the axial and transverse distributions.

$$\xi = \rho^2 - 0.5, \, q(\xi) = P(\rho). \tag{6.2}$$

Then, equation (6.1) becomes

$$h(u, w_{20}) = 2 \int_{-0.5}^{0.5} q(\xi) \exp(-j2\pi w_{20}\xi) J_0[2\pi u\sqrt{(\xi + 0.5)}] d\xi. \tag{6.3}$$

The axial behavior of the system is obtained from equation (6.3), at $u = 0$ as follows:

$$h(0, w_{20}) = \int_{-0.5}^{0.5} q(\xi) \exp(-j2\pi w_{20}\xi) d\xi = FT[q(\xi)] \tag{6.4}$$

where FT is the exact Fourier transform realized by the transformation mentioned in equation (6.2).

The transverse behavior is obtained from equation (6.3), at $w_{20} = 0$ as follows:

$$h(u, w_{20}) = \int_{-0.5}^{0.5} q(\xi) J_0[2\pi u\sqrt{(\xi + 0.5)}] d\xi. \tag{6.5}$$

These reported Eqs. (3.1–3.5) by Martinez et al. are used in the next applications using the Martinez-Corral filter.

6.2.1 Martinez-Corral Filter

The Martinez-Corral filter consists of a transparent annulus and a central clear circular aperture of area less than the area of the annulus. The irradiances are computed from Eqs. (3.4) and (3.5) to obtain these results:

$$I \text{ (axial)} = \left[\frac{\sin(\pi w_{20})}{(\pi w_{20})}\right]^2 + \mu^2 \left[\frac{\sin(\pi \mu w_{20})}{(\pi \mu w_{20})}\right]^2$$
$$- 2\mu \frac{\sin(\pi w_{20})}{(\pi w_{20})} \frac{\sin(\pi \mu w_{20})}{(\pi \mu w_{20})} \cos[2\pi w_{20}(\varepsilon - 0.5)\mu] \qquad (6.6)$$

$$I \text{ (transverse)} = \frac{1}{\pi^2 \mu^2} \{ J_1(2\pi \mu) + \sqrt{(0.5 - \varepsilon)} J_1[2\pi \mu \sqrt{(0.5 - \varepsilon)}]$$
$$- \sqrt{0.5 + (1 - \varepsilon)\mu} J_1[2\pi \mu \sqrt{0.5 + (1 - \varepsilon)\mu}]\},$$
$$q(\xi) = rect(\xi) - rect\left[\frac{\xi + (\varepsilon - 0.5)\mu}{\mu}\right] \qquad (6.7)$$

With $0.5 < \varepsilon < 1$ and $0 < \varepsilon\mu < 0.5$, μ is the obscuration parameter, and ε is the asymmetry parameter.

6.2.2 Author's Filter

The aperture under study has a definite number of black and white (B/W) annuli of a certain number of circles $N = 20$, where the center is a clear circular disk, as shown in the figure in ref. [12]. The effective pupil of this aperture is mathematically represented as follows:

$$P(\rho) = circ\left(\frac{\rho}{\rho_0}\right) + \sum_{i=1}^{N} \Delta P_i(\rho) \qquad (6.8)$$

where $\Delta P_i(\rho) = P_{2i+1}(\rho) - P_{2i}(\rho)$ is the difference between any two successive circular apertures representing an annular shape, N is the total number of circles and ρ is the radial coordinate in the aperture plane (u, v).

6.2.3 Computation of the Transverse APSF and Its Irradiance

The amplitude impulse response of the considered aperture or the amplitude point spread function (APSF) is computed by operating the Fourier transform upon the aperture represented by Eq. (3.8) to obtain the following equation:

$$h(w) = \frac{2J_1(\alpha_1 r)}{(\alpha_1 r)} + \sum_{i=1}^{N} \left\{ \frac{2J_1(\alpha_{2i+1} r)}{(\alpha_{2i+1} r)} - \frac{2J_1(\alpha_{2i} r)}{(\alpha_{2i} r)} \right\}. \qquad (6.9)$$

where $\alpha_{2i} = (2\pi/\lambda f)w_{2i}$ and $\alpha_{2i+1} = (2\pi/\lambda f)w_{2i+1}$, J_1 is the Bessel function of the 1st order and r is the radial coordinate in the Fourier plane (x, y).

In the case of non-equally spaced annuli, Eq. (6.9) is used in the computations suppressing some annuli where the obscuration parameter has $\mu = 0.7895$ giving the best transverse resolution. It corresponds to the following parameters: a central disk of radius $\rho = 1/19$ followed by the dark annulus of $\mu = 15/19 = 0.7898$, then followed by the three-remaining white–dark–white annuli. The other extreme value of $\mu = 0.0526$ will give a poorer resolution which corresponds to the following parameters: central disk of radius $\rho = 1/19$ followed by the dark annulus of $\mu = 1/19 = 0.0526$, then followed by 12 B/W annuli.

The transverse irradiance is the modulus square of the amplitude point spread function represented as follows:

$$I_{\text{trans.}}(w) = |h(w)|^2. \tag{6.10}$$

In the case of confocal imaging systems, the irradiance or the image of a point becomes:

$$I_{\text{trans.}}(w) = |h(w)|^4. \tag{6.11}$$

For an aperture composed of a very thin annular shape combined with the transparent central disk, Eq. (6.9) becomes

$$h(w) = \frac{2J_1(\alpha_1 r)}{(\alpha_1 r)} + \sum_{i=1}^{N} J_0(\alpha_{2i+1} r). \tag{6.12}$$

6.2.4 Computation of the Axial APSF and Its Irradiance

The binary filter is composed of 10 transparent annuli and 10 black annuli with a central clear disk in a B/W cascaded concentric annular arrangement, as shown in the figure in Ref. [23], is represented as follows:

$$q(\xi) = \sum_{n=1}^{N} \text{rect}(\xi - n\mu\xi_0) \tag{6.13}$$

where ξ_0 is the interval width between any two successive annuli and μ is the obscuration parameter. The 1D Fourier transform is operated upon Eq. (6.13), making use of convolution operations, to get this result for the axial APSF:

$$h(w_{20}) = \sum_{n=-N/2}^{N/2} \frac{\sin(\pi w_{20})}{(\pi w_{20})} \exp(j2\pi n\mu\xi_0 w_{20}). \tag{6.14}$$

This complex function of the axial APSF can be decomposed into real and imaginary parts as follows:

$$\text{Real}[h(w_{20})] = sinc(w_{20}) \sum_{n=-N/2}^{N/2} \cos(2\pi n\mu\xi_0 w_{20}) \tag{6.15}$$

$$\text{Im}[h(w_{20})] = sinc(w_{20}) \sum_{n=-N/2}^{N/2} \sin(2\pi n\mu\xi_0 w_{20}) \tag{6.16}$$

Where $sinc(x) = \sin(\pi x)/(\pi x)$. The corresponding axial irradiance of the ordinary optical systems is computed by taking the modulus square of the axial APSF to get this result:

$$I_{\text{axial}}(w_{20}) = sinc^2(w_{20}) \left[\sum_{n=-N/2}^{N/2} \cos(2\pi n\mu\xi_0 w_{20}) \right]^2 \tag{6.17}$$

since the second summation over the sine odd function vanished. In the case of confocal imaging, the axial irradiance is

$$I_{\text{axial}}(w_{20}) = sinc^2(w_{20}) \left[\sum_{n=-N/2}^{N/2} \cos(2\pi n\mu\xi_0 w_{20}) \right]^4. \tag{6.18}$$

This can be rewritten as follows:

$$I_{\text{axial}}(w_{20}) = sinc^4(w_{20}) \left[1 + 2 \sum_{n=-N/2}^{N/2} \cos(2\pi n\mu\xi_0 w_{20}) \right]^2. \tag{6.19}$$

6.3 Theoretical Results and Discussion

The proposed filter of multi-ring aperture of B/W concentric annuli of different obscuration parameters (μ) is used in the calculation of the axial and lateral (transverse) PSF and the corresponding irradiances. Eight graphs representing the transverse PSF are shown in Figures (6.1a–6.8a). These graphs are plotted using Eq. (6.9). The corresponding irradiances are obtained from Eq. (6.10) and represented in figures

(6.1b–6.8b). The graphs of the confocal imaging irradiances are obtained using Eq.
(6.11) and the plots are like the irradiances shown in Figs. 6.1, 6.2, 6.3, 6.4, 6.5,
6.6, 6.7 and 6.8 except that the legs of the irradiance distribution are attenuated or
suppressed. Our multi-ring arrangement is composed of a central clear disk obstructed
by a dark annulus of obscuration parameter (μ) then followed by a sandwich of B/W
annuli. Referring to the above results, it is shown that the best transverse resolution
corresponds to the following parameters: central disk of radius $\rho = 1/19$ followed
by the dark annulus of $\mu = 15/19 = 0.7898$, then followed by white–dark–white
annuli. The result corresponding to this arrangement is shown in Fig. 6.1b. In this
case, the total number of annuli $N = 19$ and the total normalized radius of the aperture is taken to be unity. It is shown that the transverse resolution is decreased as μ
is decreased reaching equally spaced concentric B/W annuli. The improvement of
transverse resolution in the Martinez filter is at the expense of the appearance of the
legs of the diffracted irradiance while only one leg of low height has appeared as
in Fig. 6.8b using our filter. Another advantage of our multi-ring arrangement lies
in its gain in intensity. The Martinez-Corral transverse PSF and its corresponding
irradiances are plotted in Fig. 6.9a, b. The axial irradiance is obtained using

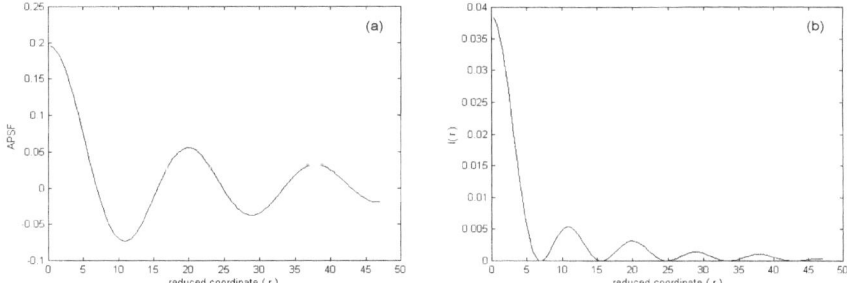

Fig. 6.1 a Normalized transverse point spread function, where the obscuration parameter $= 0.7895$.
b Normalized transverse irradiance, where the obscuration parameter $= 0.7895$

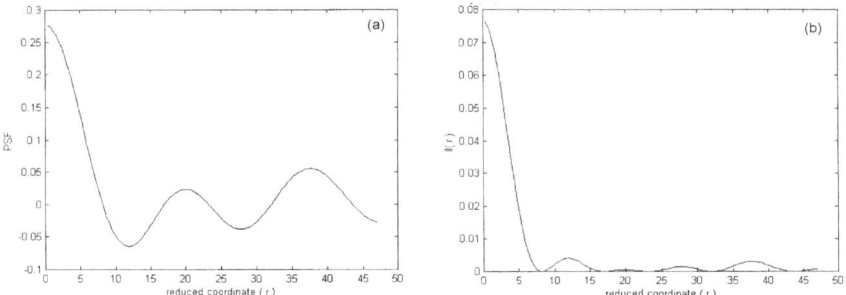

Fig. 6.2 a Normalized transverse PSF, where the obscuration parameter $= 0.6842$. **b** Normalized
transverse irradiance, where the obscuration parameter $= 0.6842$

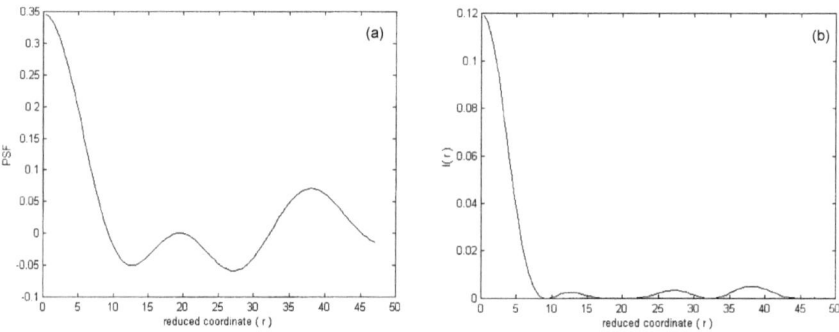

Fig. 6.3 a Normalized transverse PSF, where the obscuration parameter $= 0.5789$. **b** Normalized transverse irradiance, where the obscuration parameter $= 0.5789$

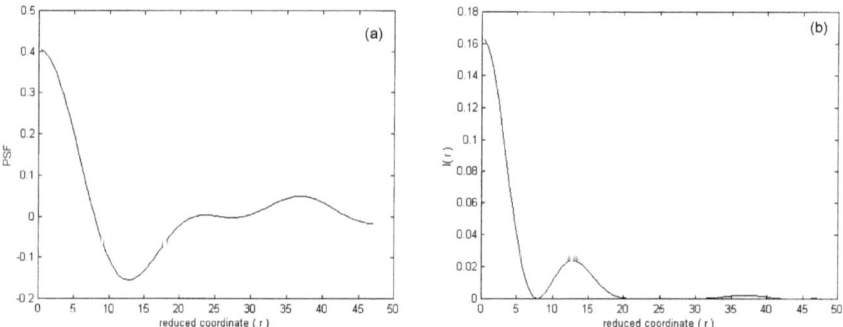

Fig. 6.4 a Normalized transverse PSF, where the obscuration parameter $= 0.4737$. **b** Normalized transverse irradiance, where the obscuration parameter $= 0.4737$

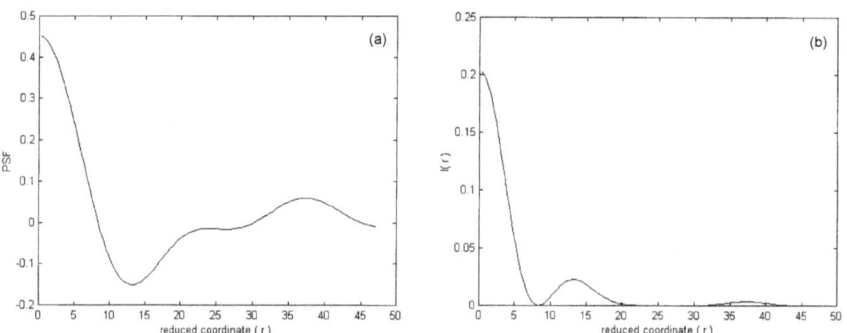

Fig. 6.5 a Normalized transverse PSF, where the obscuration parameter $= 0.3684$. **b** Normalized transverse irradiance, where the obscuration parameter $= 0.3684$

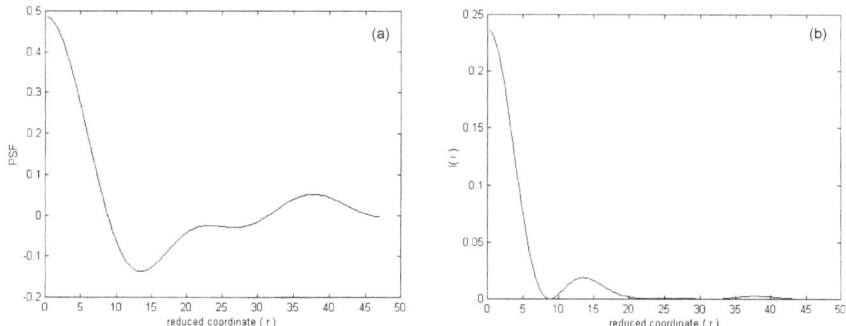

Fig. 6.6 **a** Normalized transverse PSF, where the obscuration parameter = 0.2632. **b** Normalized transverse irradiance, where the obscuration parameter = 0.2632

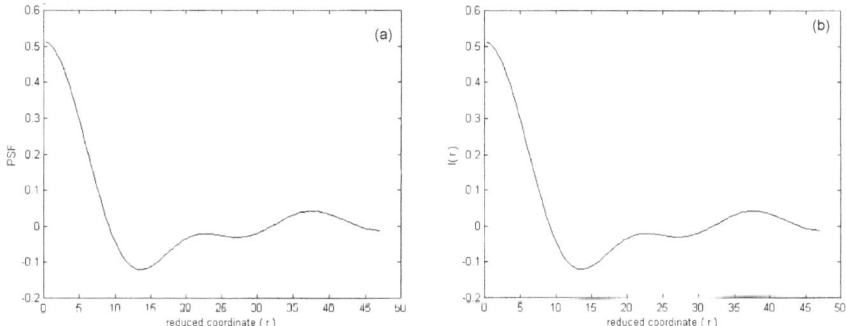

Fig. 6.7 **a** Normalized transverse PSF, where the obscuration parameter = 0.1579. **b** Normalized transverse irradiance, where the obscuration parameter = 0.1579

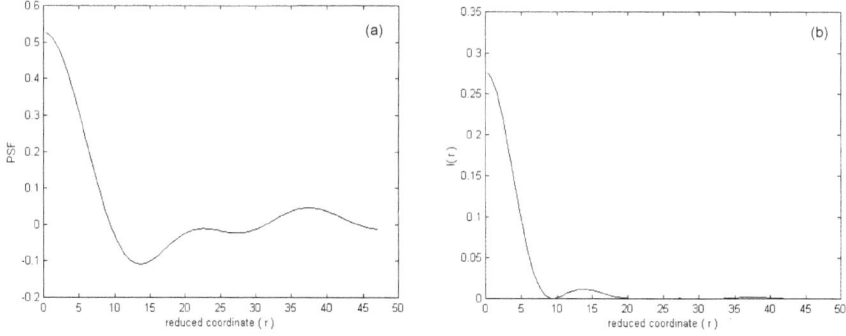

Fig. 6.8 **a** Normalized transverse PSF, where the obscuration parameter = 0.0526. **b** Normalized transverse irradiance, where the obscuration parameter = 0.0526

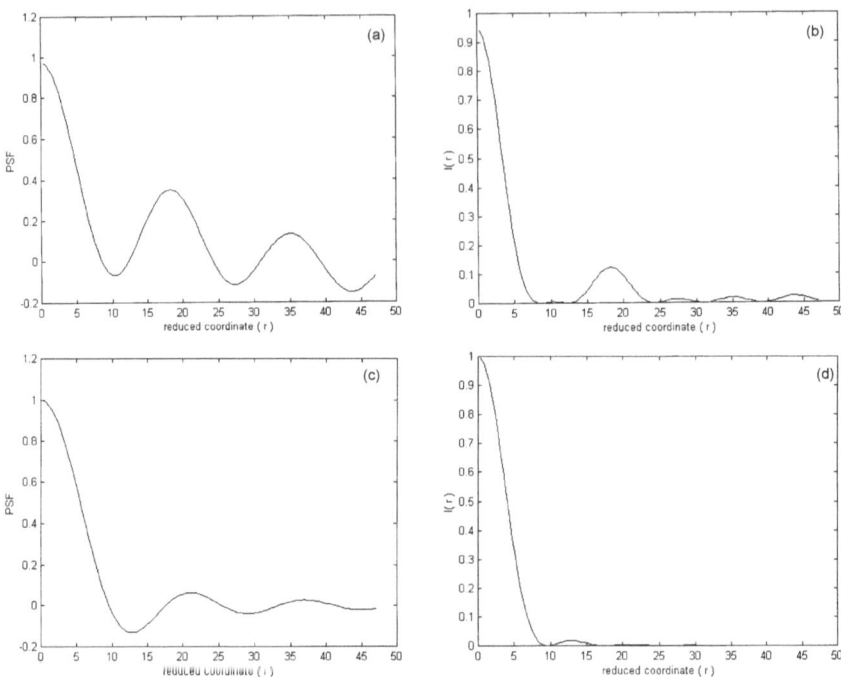

Fig. 6.9 **a** Normalized transverse PSF for the Martinez filter. **b** Normalized transverse irradiance for Martinez filter. **c** Normalized transverse PSF for circular aperture. **d** Normalized transverse irradiance for circular aperture

Equation (6.17) and represented as in Fig. 6.10a. Another plot corresponding to the Martinez filter is made using Eq. (6.6) where $\mu = 0.7$ as shown in Fig. 6.10b. Also, a separate curve of irradiance distribution for circular aperture appeared as in Fig. 6.10c. A set of axial irradiances for the different filters and circular aperture are combined and plotted as shown in Fig. 6.10d, where the discontinuous curve is set for circular aperture and the curve corresponding to our filter has no legs as compared with the Martinez curve which has a dominant distribution outside the central band. It is shown that our axial irradiance curve lies between the Martinez curve and the circular aperture discontinuous curve. Referring to the results shown in Fig. 6.10d our filter has a resolution better than that obtained for the circular curve and less than the Martinez resolution. The advantage of our filter when compared with the Martinez filter lies in its suppression of the legs of the axial irradiance and the reasonable gain obtained from the detected intensity. It is shown, referring to the results of the axial PSF for different values of μ, as shown in Fig. 6.11a–d, that as the obscurity parameter is increased the axial resolution is improved. As μ is decreased the axial resolution becomes poorer and the legs of the diffraction pattern will appear. Hence, it is recommended to take the greater value of the obscuration parameter to get better resolution and equally suppress the legs of the diffraction pattern.

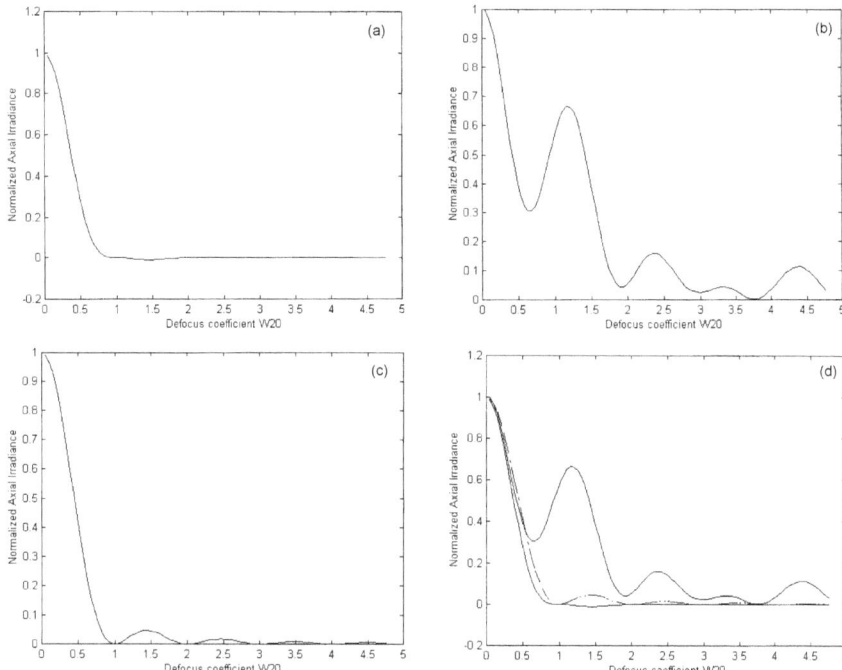

Fig. 6.10   a Normalized axial irradiance for B/W concentric annuli. **b** Normalized axial irradiance for the Martinez filter. **c** Normalized axial irradiance for clear circular aperture. **d** Normalized axial irradiance for B/W annuli, Martinez filter, and circular aperture

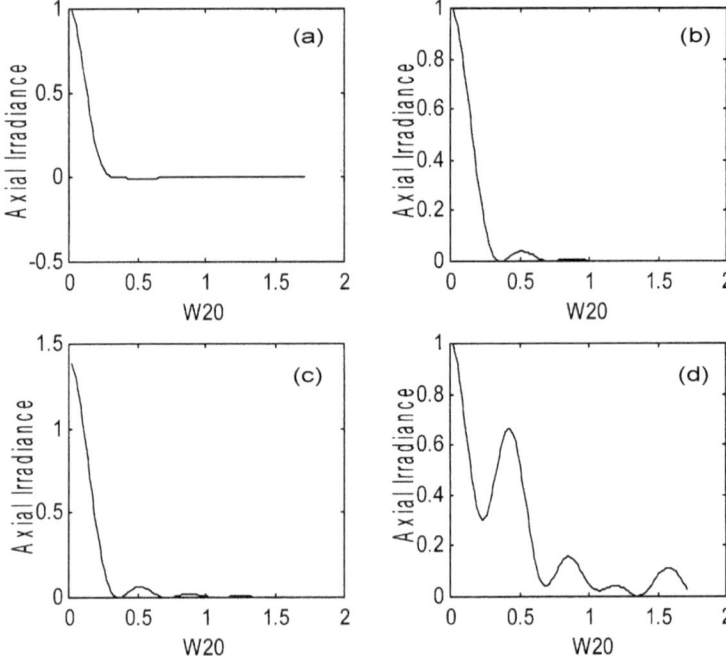

Fig. 6.11. a Author's axial irradiance, $\mu = 1.0$. **b** *I*-axial, $\mu = 0.8$. **c** *I* axial, $\mu = 0.5$ **d** Martinez axial irradiance

6.4   Conclusion

The model of the multi-ring aperture of B/W concentric annuli is presented. We have computed both the transverse and axial PSF and the corresponding irradiances. The obtained results are compared with the results of the Martinez-Corral annular obstruction filter and also compared with the clear circular aperture. It is concluded that the obtained transverse resolution is better than that obtained for circular aperture while it is less than that obtained for the Martinez filter. It is found that the transverse resolution is dependent upon the obscuration parameter μ and is improved for greater values of μ in the case of our filter. Hence, further improvement of resolution may be attained in the case of non-equally spaced annuli which is dependent upon the obscuration parameter μ. The axial resolution corresponding to our filter lies between the Martinez resolution and circular aperture resolution. In the case of the Martinez filter the legs of the irradiance distribution appeared while in the case of our filter, complete suppression of the legs occurred for $\mu = 1.0$ making our amplitude filter better than the Martinez filter.

References

1. C.J.R. Sheppard, A. Choudhary, Image formation in the scanning microscope. Opt. Acta **24**, 1051–1073 (1977)
2. C.J.R. Sheppard, T. Wilson, J. Microscopy **114**, 179 (1978)
3. T. Wilson, Imaging properties and applications of scanning optical microscopes. Appl. Phys. **22**, 119–128 (1980)
4. C.J.R. Sheppard, The use of lenses with annular aperture in scanning optical microscopy. Optik **48**, 329–334 (1977)
5. C.J.R. Sheppard, T. Wilson, Image formation in scanning microscopes with partially coherent source and detector. Opt. Acta **25**, 315–325 (1978)
6. A.M. Hamed, Optimization of spatial coherence in confocal optical systems. Opt. Laser Tech. **22**, 137–139 (1990). https://doi.org/10.1016/0030-3992(90)90024-X
7. A.M. Hamed, A study on spatial coherence using quadratic radially distributed apertures (application to confocal imaging). Opt. Laser Tech. **29**, 93–95 (1997)
8. J.D. Gaskill, *Linear systems, fourier transform and optics* (Wiley, New York, 1978)
9. J.W. Goodman, *Introduction to fourier optics and holography* (McGraw-Hill Co, New York, 1968)
10. R.N. Bracewell, *Fourier transform and its applications* (McGraw-Hill Co, New York, 1966)
11. J.J. Clair, A.M. Hamed, Theoretical studies on optical coherent microscope. Optik **64**, 133–141 (1983)
12. A.M. Hamed, J.J. Clair, Image and super-resolution in optical coherent microscopes. Optik **64**, 277–284 (1983)
13. A.M. Hamed, J.J. Clair, Studies on optical properties of confocal scanning optical microscope using pupils with radially transmission distribution. Optik **65**, 209–218 (1983)
14. A.M. Hamed, Resolution and contrast in confocal optical scanning microscope. Opt. Laser Tech. **16**, 93–96 (1984)
15. A.M. Hamed, A study on amplitude modulation and an application on confocal imaging. Optik **107**, 161–164 (1998)
16. C.J.R. Sheppard, Z.S. Hegedus, Axial behavior of pupil-plane filters. J. Opt. Soc. Am. A **5**, 643–647 (1988)
17. I.J. Cox, C.J.R. Sheppard, T. Wilson, Reappraisal of arrays of concentric annuli as super resolving filters. J. Opt. Soc. Am. **72**, 1287–1291 (1982)
18. Z.S. Hegedus, V. Sarafis, Supper resolving filters in confocal scanned imaging systems. J. Opt. Soc. Am. A **3**, 1892–1896 (1986)
19. T.R.M. Sales, G.M. Morris, Diffractive super-resolution elements. J. Opt. Soc. Am. A **14**, 1637–1646 (1997)
20. T.R.M. Sales, G.M. Morris, Axial super-resolution with phase-only pupil filters. Opt. Commun. **156**, 227–230 (1998)
21. C.J.R. Sheppard, G. Calvert, M. Wheatland, Focal distribution for super resolving Toraldo filters. J. Opt. Soc. Am. **A15**, 849–856 (1998)
22. H. Wang, F. Gan, High focal depth with a pure phase apodizer. Appl. Opt. **40**, 5658–5662 (2001)
23. M. Martinez-Corral, M.T. Caballero et al., Tailoring the axial shape of the point spread function using the TOLARDO concept. Opt. Express **10**, 98–103 (2002)
24. C.M. Blanca, W. Hell Stefan, Axial super-resolution with ultrahigh aperture lenses. Opt. Express **10**, 893–898 (2002)
25. L. Cheng, G.G. Siu, Asymmetric apodization. Meas. Sci. Technol. **2**, 198 (1991)
26. G.G. Siu, L. Cheng et al., Improved side-lobe suppression in asymmetric apodization. J. Phys. D **27**, 459 (1994)
27. A.M. Hamed, Computation of the lateral and axial point spread function in confocal imaging systems using binary amplitude mask. J. Phys. **66**, 1037–1048 (2006)
28. O. Hanselman, B. Littlefield, Mastering MATLAB 5 edited by M Horton (Upper Saddle River, New Jersey 07458, 1998) Ch. 22, p. 238

Chapter 7
Exenteration Errors Combined with Wavefront Aberration

7.1 Introduction

The basic concept of a Coherent Scanning Microscope (C.S.M.) was given by Minsky [1] in his patent application, while Davidovits et al. [2] built this microscope using direct laser illumination. However, spatial coherence conditions have not been considered in this early work.

NOMARSKI [3] has given the right formula for the depth of field of C.S.M. Following him, Sheppard and his collaborators have contributed numerous papers [4–9] to the development of his new technique of microscopic imaging (named confocal imaging). Hamed et al. [10–18] have proposed novel methods, using amplitude modulation techniques, to improve the resulting impulse response of the C.S.M.

A pinhole detector is necessary to obtain a coherent image. Practically all imaging systems exhibit a certain degree of coherence, so the illumination is always partially coherent. Hence, it is necessary to choose the pinhole diameter small enough to gain sufficient output energy, i.e., compromising between the illuminating conditions and the pinhole diameter of the detector must be found to enhance the image contrast.

The confocal arrangement of the objectives L_1 and L_2 combined with the scanned object are the basic elements of the C.S.M. as shown in Fig. 7.1. In this schematic diagram, the scanning of both the object and electron beam emitted from the cathode ray monitor simultaneously will ensure the construction of the temporal image. Unfortunately, the C.S.M. either working in transmission or at reflection will give a weak output signal that needs amplification before its detection. For example, the heterodyne technique of detection may be useful in this case to improve signal-to-noise ratio S/N. We pay attention that the weak output signal simultaneously with the required high precision of object mechanical scanning will make this microscope much more difficult its realization than the well-known classical optical microscopes.

The purpose of this study is to show the influence of the exenteration of the objectives on the resulting impulse response obtained by the C.S.M. A misaligned optical system of the C.S.M. if only the 2nd objective exhibits a lateral shift is shown

© The Author(s), under exclusive license to Springer Nature Switzerland AG 2025   81
A. M. Hamed, *Studies on the Confocal Laser Microscope*,
SpringerBriefs in Applied Sciences and Technology,
https://doi.org/10.1007/978-3-031-87275-4_7

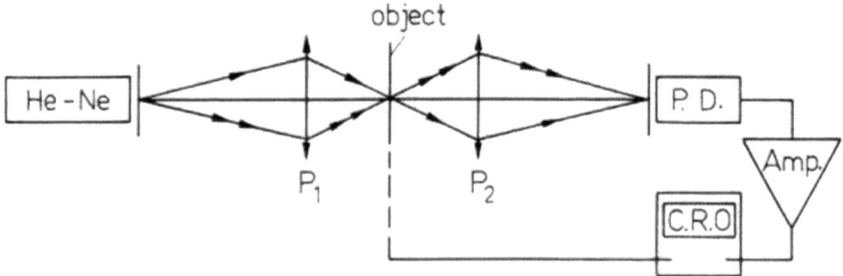

Fig. 7.1 Setup of the aligned optical coherent scanning microscope provided with two confocal objectives

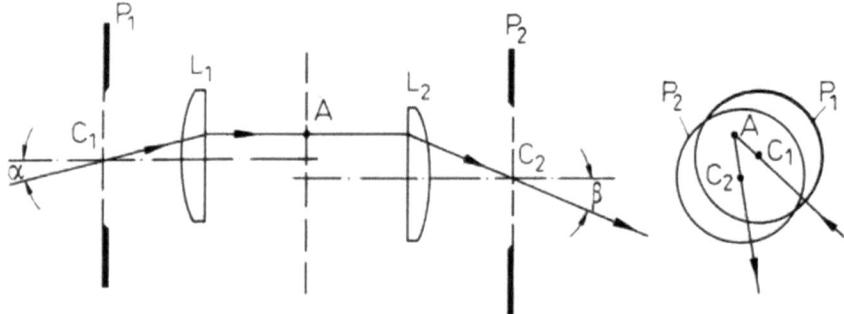

Fig. 7.2 Setup of a misaligned optical system with laterally displaced objectives

in Fig. 7.2. Another misaligned optical system in which L_2 is tilted and shifted concerning the optical axis of the aligned objective L_1 is shown in Fig. 7.4. These serious problems of exenteration will be analyzed in the following section taking into consideration the classical wavefront aberration.

In this chapter, I calculated the exenteration errors combined with the wavefront aberration using uniform circular apertures. I extended the studies on exenteration errors using linear and quadratic apertures.

7.2 Mathematical Formulation of the Problem

The intensity distribution $I(w)$ of a coherent scanning laser microscope (CSLM) shown in Fig. 7.1 is calculated from the modulus square of the following convolution integral represented symbolically in Eq. (7.1):

$$I(w) = |g(w) \otimes h_r(w)|^2 \tag{7.1}$$

$g(w)$ is the complex amplitude transmitted from the object, and $h_r(w) = h_1(w) \cdot h_2(w)$ is the resulting impulse response. h_1 is given for the first objective lens, while h_2 is given for the second objective. $\otimes$: symbol for the convolution.

7.2.1 Using Circular Apertures

7.2.1.1 Effect of Exenteration Errors

Since it is easy to align the first objective of the CSLM while this is not the case for the second objective, we assume that the latter objective is subjected to tilting and a lateral shift concerning the first aligned objective. Under the above conditions, the complex amplitude transmitted from the first objective for any oblique ray is:

$$P_1(\rho; \alpha) = P_{01} \exp(jk\rho \sin \alpha), \quad |\rho/\rho_{01}| \leq 1 \tag{7.2}$$

and that corresponding to the 2nd shifted objective is:

$$P_2(\rho; \beta, \rho_d) = P_{02}(\rho; \rho_d) \exp(jk\rho \sin \beta), \quad |(\rho \pm \rho_d)/\rho_{02}| \leq 1 \tag{7.3}$$

where α and β are the tilting angles of the corresponding objectives L_1 and L_2, respectively, and ρ_d is the lateral shift introduced w.r.t. the optical axis. The sign plus or minus appears in Eq. (7.3) to determine the direction of the displacement given to L_2 to the optical axis, as shown in Fig. 7.2.

The point spread function is obtained by performing the Fourier transformation on Eq. (7.2). After applying convolution operations, we obtain the following result for the PSF for the 1st objective [19–21]:

$$h_1(w; \alpha) = 2\pi\rho_{01}^2 \frac{J_1(w)}{w} \otimes \delta(r - f \sin \alpha) \tag{7.4}$$

It is rewritten as follows:

$$h_1(w'; \alpha) = 2\pi\rho_{01}^2 \left[\frac{J_1(w')}{w'} \right] \tag{7.5}$$

The new reduced coordinate w' is related to the ordinary reduced coordinate w through this equality:
$w' = (w/r)(r - f \sin \alpha); \ w = \left(\frac{2\pi}{\lambda f} \right)\rho_{01}r$. The relative amount of shift $\left| \frac{\delta w}{w} \right| = \frac{f \sin \alpha}{r}$.

The calculation of the PSF for the 2nd objective is not straightforward. Hence, before performing the transformation, it is helpful to change the radial limits of the integration to $\rho_{min} = \pm\rho_d$ and $\rho_{max} = \rho_{02} \pm \rho_d$, We obtain the following result:

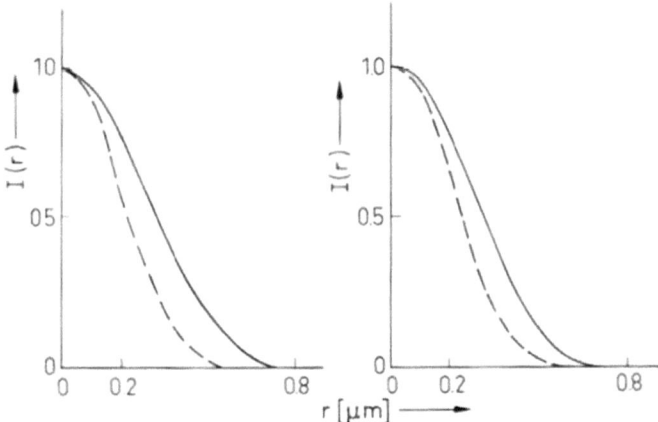

Fig. 7.3 Intensity response of the CSLM, where the 2nd objective is laterally shifted by an amount of ρ_d, wher $\rho_d = 100$ μm in Fig. 7.3a and $\rho_d = 500$ μm in Fig. 7.3b. The solid line is for Classical M. while the discontinuous line stands for the CSLM.

$$h_2(w; \beta, \rho_d) = 2\pi(\rho_{02} \pm \rho_d)^2 \left\{ \left[\frac{J_1(w_u)}{(w_u)} \right] - \tau^2 \left[\frac{J_1(w_L)}{(w_L)} \right] \right\} \qquad (7.6)$$

where $\tau = \rho_d/(\rho_{02} \pm \rho_d)$, $w_u = \left(\frac{2\pi}{\lambda f} \right)(\rho_{02} \pm \rho_d)(r - f \sin \beta)$, and $w_L = \left(\frac{2\pi}{\lambda f} \right)\rho_d(r - f \sin \beta)$

The upper limit of integration w_u is less than w by $\left(1 \pm \frac{\rho_d}{\rho_{02}} \right)\left(1 - \frac{f}{r} \sin \beta \right)$, and the lower limit w_L is less than w by $\left(\frac{\rho_d}{\rho_{02}} \right)\left(1 - \frac{f}{r} \sin \beta \right)$.

The resulting impulse response is deduced from the product of Eqs. (7.5) and (7.6) as:

$$h_r(w; \alpha, \beta, \rho_d) = 4\pi^2 \rho_{01}^2 (\rho_{02} \pm \rho_d)^2 \left[\frac{J_1(w')}{w'} \right] \left\{ \left[\frac{J_1(w_u)}{(w_u)} \right] - \tau^2 \left[\frac{J_1(w_L)}{(w_L)} \right] \right\}$$

$$(7.7)$$

7.2.1.2   Effect of Exenteration Combined with Aberration

We calculate the intensity of the impulse response in the general case where the objectives of the CSLM are influenced by a 3rd order coma in addition to the exenteration errors, as shown in Fig. 7.4. These defects are harmful to the resolution of the microscope. We assume that the 2nd objective is influenced by tilting and lateral shifting combined with a wavefront aberration. The transmitted amplitude is:

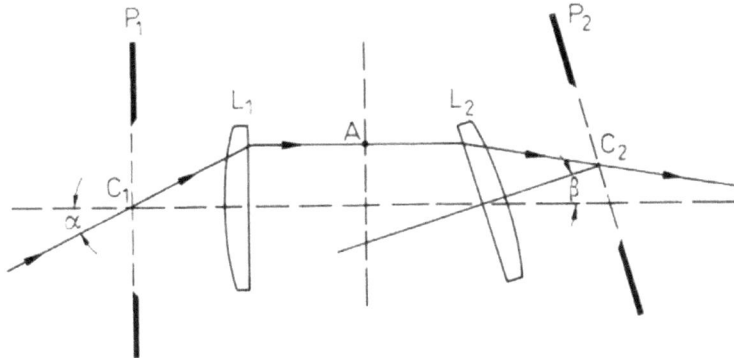

Fig. 7.4 Setup of a misaligned optical system providing objectives with orthogonal tilting combined with lateral shifting

$$P_2(\rho; \rho_d, \beta, c_1) = \underbrace{P_{02}(\rho; \rho_d)}_{|\leftarrow \text{shift} \rightarrow|} \cdot \underbrace{\exp(jk\rho \sin \beta)}_{\leftarrow \text{tilting} \rightarrow} \cdot \underbrace{\exp(jk\Delta(\rho, \phi))}_{|\leftarrow \text{aberration} \rightarrow|} \qquad (7.8)$$

In the case of a small phase k, we can develop it in a series form [19, 20] as:

$$\Delta(\rho, \phi) = \rho^2 d + \sum_{n=2}^{N} s_n \rho^{2n} + \sum_{n=1}^{N} c_n \rho^{2n+1} \cos \phi + a\rho^2 \cos 2\phi \qquad (7.9)$$

where d is the defect of the focus coefficient, s is the spherical aberration, c: is the coefficient of coma, and a: is the coefficient of astigmatism.

Retaining only the 3rd order coma, Eq. (7.9) becomes:

$$\Delta(\rho, \phi) = c_1 \rho^3 \cos \phi \qquad (7.10)$$

Where $\emptyset$ are the azimuthal coordinates in the objective plane.

If the deviation of the wavefront does not exceed $\lambda/4$, then we retain from Eq. (7.10). The terms up to the 2nd order:

$$\exp(ik\Delta(\rho, \phi)) \cong 1 + jk\Delta(\rho, \phi) - \frac{k^2}{2}\Delta^2(\rho, \phi) = 1 + jkc_1\rho^3 \cos \phi - \frac{k^2}{2}c_1^2\rho^6 \cos^2 \phi \qquad (7.11)$$

From Eq. (6.8) and Eq. (6.11), we write the aberration pupil function as:

$$P_2(\rho; \rho_d, \beta, c_1) = P_{02}(\rho; \rho_d). \exp(jk\rho \sin \beta).[1 + jkc_1\rho^3 \cos \phi - \frac{k^2}{2}c_1^2\rho^6 \cos^2 \phi] \qquad (7.12)$$

The Fourier transformation is applied to Eq. (7.12) to obtain the PSF as:

$$h_2(w; \rho_d, \beta, c_1) = F.T.\{P_{02}(\rho; \rho_d)\} \otimes \delta(r - f \sin \beta)$$

$$+ jkc_1 F.T.\{P_{02}(\rho; \rho_d) \cdot \rho^3 \cos \phi\} \otimes \delta(r - f \sin \beta) - \frac{k^2}{4} \cdot$$

$$c_1^2 F.T.\{P_{02}(\rho; \rho_d) \cdot \rho^6 (1 + \cos 2\phi)\} \otimes \delta(r - f \sin \beta) \qquad (7.13)$$

Making use of the properties of the Bessel function $J_n(x) = \frac{i^{-n}}{2} \int_0^{2\pi} e^{ix \cos \phi} e^{in\phi} d\phi$,
we find that:

F.T.$\{P_{02}(\rho; \rho_d) . \rho^3 \cos \phi\} = 0$, and F.T.$\{P_{02}(\rho; \rho_d) \cdot \rho^6 \cos 2\phi\} = 0$.
We have this result for the PSF:

$$h_2(w; \rho_d, \beta, c_1) = 2(\rho_{02} \pm \rho_d)^2 A(w; \beta, \rho_d) - 2(\rho_{02} \pm \rho_d)^8 [B(w; \beta, ; \rho_d, c_1)$$

$$- 6C(w; \beta, ; \rho_d, c_1) + 24D(w; \beta, ; \rho_d, c_1) - 48E(w; \beta, ; \rho_d, c_1)]$$
$$(7.14)$$

With

$$A = \left[\frac{J_1(w_u)}{(w_u)}\right] - \tau^2 \left[\frac{J_1(w_L)}{(w_L)}\right], \qquad B = \frac{k^2}{4}c_1^2 \left\{\left[\frac{J_1(w_u)}{(w_u)}\right] - \tau^8 \left[\frac{J_1(w_L)}{(w_L)}\right]\right\},$$

$$C = \frac{k^2}{4}c_1^2 \left\{\left[\frac{J_2(w_u)}{(w_u^2)}\right] - \tau^8 \left[\frac{J_2(w_L)}{(w_L^2)}\right]\right\}, \quad D = \frac{k^2}{4}c_1^2 \left\{\left[\frac{J_3(w_u)}{(w_u^3)}\right] - \tau^8 \left[\frac{J_3(w_L)}{(w_L^3)}\right]\right\},$$

$$E = \frac{k^2}{4}c_1^2 \left\{\left[\frac{J_4(w_u)}{(w_u^4)}\right] - \tau^8 \left[\frac{J_4(w_L)}{(w_L^4)}\right]\right\}$$

The resulting impulse response and its intensity are calculated from Eqs. (7.5)
and (7.14) as follows:

$$h_r = h_1 \cdot h_2 \text{ and } I = |h_r(w; \alpha, \beta, \rho_d, c_1)|^2 \qquad (7.15)$$

In the case of perfect alignment, of the optical system ($\beta = 0$, $\rho_d = 0$), we obtain
a result valid for objective suffering only from a simple coma where $\tau = 0$:

$$h_2(w; c_1) = 2\pi \left(1 - \frac{k^2}{4}c_1^2\right)\left[\frac{J_1(w)}{w}\right] + (3\pi k^2 c_1^2)\left[\frac{J_2(w)}{w^2} - 4\frac{J_3(w)}{w^3} + 8\frac{J_4(w)}{w^4}\right]$$
$$(7.16)$$

The pupil radius is taken to be unity, $\rho_{02} = 1$.

In this case, the intensity response is deduced from Eqs. (7.5) and (7.16). For free
aberration and perfectly aligned confocal systems, one obtains the result given by
Sheppard as [4]:

$$I(w) = \left[\frac{J_1(w)}{w}\right]^4 \qquad (7.17)$$

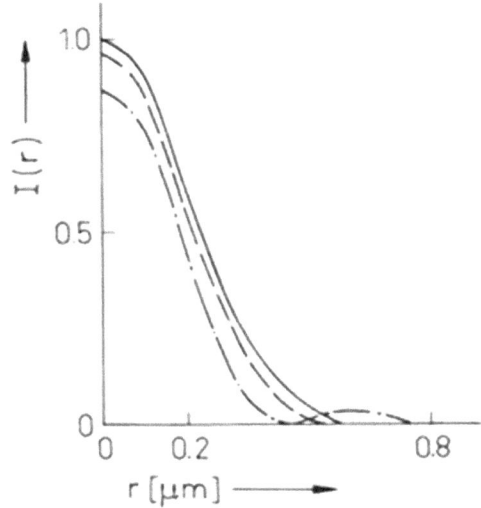

Fig. 7.5 Intensity response of a CSLM where the second objective is influenced by a third-order coma. Coefficient of coma $c_1 = 0$, 0.05, and 0.1. The solid curve represents no coma $c_1 = 0$, the discontinuous curve (–) represents $c_1 = 0.05$, and the discontinuous dotted curve represents $c_1 = 0.1$

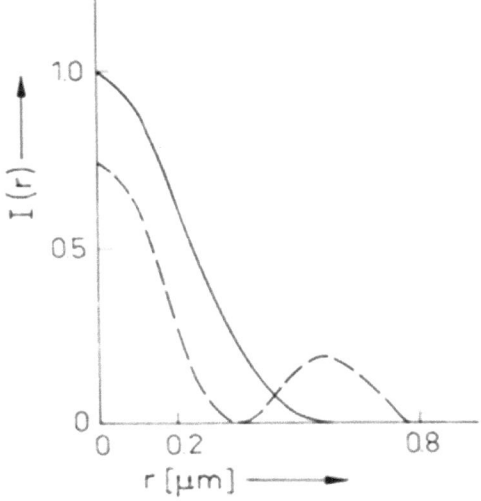

Fig. 7.6 Intensity response of a C.S.M., where the second objective is influenced by a coma of $c_1 = 0.15$, represented by the discontinuous line, while the solid line is given for $c_1 = 0$ for comparison

Some results corresponding to objectives influenced by coma are shown in Figs. 7.5 and 7.6.

7.2.2 Misaligned Linear and Quadratic Apertures

We investigate the PSF in the case of misaligned linear and quadratic apertures.

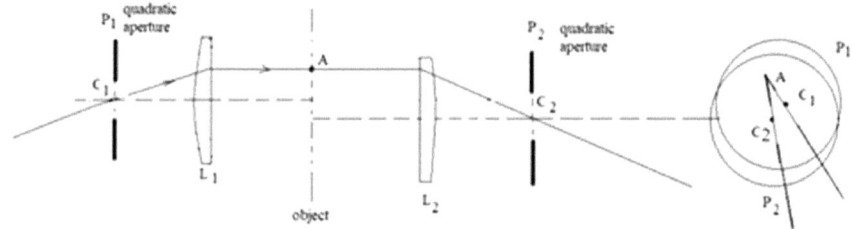

Fig. 7.7 Setup of a misaligned optical system with laterally displaced objectives using quadratic apertures in the CSLM

The CSLM under investigation is shown in Fig. 7.7, where the first objective is aligned perfectly and the second objective is displaced laterally by a radial amount of ρ_d. In this case, we write the complex amplitude from the first aligned aperture of the linear distribution as follows:

$$P_1(\rho; \alpha) = \rho \exp[jk \sin(\alpha)]; \quad \left|\frac{\rho}{\rho_{01}}\right| \leq 1 \tag{7.18}$$

where $j = \sqrt{-1}$, $k = 2\pi/\lambda$, and ρ is the radial coordinate in the aperture.

The complex amplitude for the second displaced linear aperture is written as follows:

$$P_2(\rho; \beta, \rho_d) = (\rho - \rho_d) \exp[jk\rho \sin(\beta)]; \quad \left|\frac{\rho}{\rho_{02}}\right| \leq 1 \tag{7.19}$$

where α and β are the tilting angles of the corresponding objectives L_1 and L_2, respectively.

Similar equations are written, for the quadratic apertures where the first objective is aligned while the second is laterally displaced, as follows:

$$P_3(\rho; \alpha) = \rho^2 \exp[jk\rho \sin(\alpha)]; \quad \left|\frac{\rho}{\rho_{01}}\right| \leq 1 \tag{7.20}$$

$$P_4(\rho; \beta, \rho_d) = (\rho - \rho_d)^2 \exp[jk\rho \sin(\beta)]; \quad \left|\frac{\rho}{\rho_{02}}\right| \leq 1 \tag{7.21}$$

Using Fourier transform operations and convolution products, we can obtain the PSF corresponding to each aperture as follows:

The PSF for the first aligned linear aperture, represented by Eq. (7.18), is solved to give the following result:

$$h_1(r) = \text{F.T.}[P_1(\rho; \alpha)]$$
$$= \text{F.T.}\{\rho \exp[jk\rho \sin(\alpha)]\}$$
$$= \text{F.T.}\{\rho\} \otimes F.T.\{ \exp[jk\rho \sin(\alpha)]$$
$$= \text{F.T.}\{\rho\}\delta(r - f \sin\alpha) \tag{7.22}$$

Since

$$\text{F.T.}\{\rho\} = 2\pi \left[\frac{J_1(W)}{W} + \frac{J_0(W)}{W^2} - 2 \sum_i \frac{J_i(W)}{W^3} \right]; \text{ref.[5]} \tag{7.23}$$

From Eqs. (7.22) and (7.23), we finally obtain h_1 for the 1st aligned linear aperture as follows:

$$h_1(W; \alpha) = 2 \left[\frac{J_1(W')}{W'} + \frac{J_0(W')}{W'^2} - 2 \sum_i \frac{J_i(W')}{W'^3} \right] \tag{7.24}$$

where $w' = \frac{2\pi\rho_{01}}{\lambda f}(r - f \sin\alpha)$ is the reduced coordinate in the Fourier plane.

The PSF for the displaced linear aperture represented by Eq. (7.19) is solved by operating the F.T. to give the following result:

$$h_2(r) = \text{F.T.}\{(\rho - \rho_d) \exp[jk\rho \sin(\beta)]\}$$
$$= \text{F.T.}\{(\rho) \exp[jk \sin(\beta)]\} - \text{F.T} \{(\rho_d) \exp[jk\rho \sin(\beta)]\}$$
$$= \text{F.T.}[(\rho)] \otimes F.T.\{\exp[jk\rho \sin(\beta)]\} - (\rho_d)\text{F.T.}\{\exp[jk \sin(\beta)]\}$$
$$= [\text{F.T.}(\rho)] \otimes \delta(r - f \sin\beta) - (\rho_d)\delta(r - f \sin\beta)$$
$$= \left\{ 2 \left[\frac{J_1(W)}{W} + \frac{J_0(W)}{W^2} - 2 \sum_i \frac{J_i(W)}{W^3} \right] \right\}$$
$$\otimes \delta(r - f \sin\beta) - (\rho_d)\delta(r - f \sin\beta) \tag{7.25}$$

We finally obtain the following result for the PSF [20, 21]:

$$h_2(W; \beta, \rho_d) = 2\pi \left[\frac{J_1(W'')}{W''} + \frac{J_0(W'')}{(W''^2)} - 2 \sum_i \frac{J_i(W'')}{W''^3} \right] - (\rho_d)\delta(r - f \sin\beta) \tag{7.26}$$

With $W'' = \frac{2\pi\rho_{02}}{\lambda f}(r - f \sin\beta)$ is the reduced coordinate in the Fourier plane.

Hence, the resultant PSF for the CSLM provided with linear apertures one aligned while the second misaligned is computed from the following equation:

$$h_r(\text{linear apert.}) = h_1(W; \alpha) \cdot h_2(W; \beta, \rho_d) \tag{7.27}$$

In the case of the quadratic aperture, the PSF for the first aligned quadratic aperture, is solved to give the following result:

$$h_3(r) = \text{F.T.}\{\rho^2 \exp[jk\rho \sin(\alpha)]\} = \text{F.T.}\{\rho^2\} \otimes \text{F.T.}\{\exp[jk\rho \sin(\alpha)]\}$$
$$= \text{F.T.}\{\rho^2\} \otimes \delta(r - f \sin\alpha) \tag{7.28}$$

Since

$$\text{F.T.}\{\rho^2\} = 2\pi \left[\frac{J_1(W)}{W} - 2\frac{J_2(W)}{W^2} \right]; \text{ref.}[6] \tag{7.29}$$

From equations. (7.28) and (7.29), we finally obtain h_3 for the 1st aligned quadratic aperture as follows:

$$h_3(W; \alpha) = 2\pi \left[\frac{J_1(W')}{W'} - 2\frac{J_2(W')}{W'^2} \right] \tag{7.30}$$

where W' is defined as the linear aperture.

The PSF for the displaced quadratic aperture represented by Eq. (7.21) is solved by operating the F.T. to give the following result:

$$h_4(r) = \text{F.T.}\{(\rho - \rho_d)^2 \exp[jk\rho \sin(\beta)]\}$$
$$= \text{F.T.}\{(\rho^2 - 2\rho\rho_d + \rho_d^2) \exp[jk\rho \sin(\beta)]\}$$
$$= \text{F.T.}\{\rho^2 \exp[jk \sin\rho(\beta)]\} - 2\rho_d \text{F.T.}\{\rho \exp[jk\rho \sin(\beta)]\}$$
$$+ \rho_d^2 \text{F.T.}\{\exp[jk\rho \sin(\beta)]\} \tag{7.31}$$

The 1st term is given for the aligned quadratic aperture represented in Eq. (7.30), the 2nd term is given in Eq. (7.24) multiplied by a constant value $2\rho_d$, while the 3rd—term is a simply shifted Dirac-delta function located at $f \sin\beta$ in the Fourier plane of radial coordinate r and multiplied by the shift squared ρ_d^2.

Finally, the PSF for the 2nd shifted quadratic aperture is obtained as follows:

$$h_4(W; \beta, \rho_d)$$
$$= 2\pi \left\{ \left[\frac{J_1(W'')}{W''} - 2\frac{J_2(W'')}{W''^2} \right] - 2\rho_d \left[\frac{J_1(W'')}{W''} + \frac{J_0(W'')}{W''^2} - 2\sum_i \frac{J_i(W'')}{W''^3} \right] \right\}$$
$$+ \rho_d^2 \delta(r - f \sin\beta) \tag{7.32}$$

where W'' is defined as in the case of a shifted linear aperture.

The RPSF for the CSLM provided with quadratic apertures one aligned while the second misaligned is computed from the following equation:

$$h_r(\text{quadratic apert.}) = h_3(W; \alpha) \cdot h_4(W; \beta, \rho_d) \tag{7.33}$$

For the incident collimated parallel rays on the apertures, the inclination angles α and β are equal to zero. Hence, the PSF represented by Eqs. (7.24), (7.26), (7.30), and (7.32) becomes:

$$h_1(W) = 2\pi \left[\frac{J_1(W)}{W} + \frac{J_0(W)}{W^2} - 2 \sum_i \frac{J_i(W)}{W^3} \right]; \quad \text{1st aligned linear aperture}$$

(7.34)

$$h_2(W; \rho_d) = 2\pi \left[\frac{J_1(W)}{W} + \frac{J_0(W)}{W^2} - 2 \sum_i \frac{J_i(W)}{W^3} \right]$$
$$- (\rho_d)\delta(r); \quad \text{2nd shifted linear aperture}$$

(7.35)

$$h_3(W) = 2\pi \left[\frac{J_1(W)}{W} - 2\frac{J_2(W)}{W^2} \right]; \quad \text{1st aligned quadratic aperture}$$

(7.36)

$$h_4(W; \rho_d) = 2\pi \left\{ \left[\frac{J_1(W)}{W} - 2\frac{J_2(W)}{W^2} \right] - 2\rho_d \left[\frac{J_1(W)}{W} + \frac{J_0(W)}{W^2} - 2 \sum_i \frac{J_i(W)}{W^3} \right] \right\}$$
$$+ \rho_d^2 \delta(r); \quad \text{2nd shifted quadratic aperture}$$

(7.37)

From equations. (7.23), and (7.26), we obtain the following result for the effective PSF:

$$\begin{aligned}
h_{\text{eff}}(\text{linear apert.}) &= h_1(W) \cdot h_2(W; \rho_d) \\
&= h_1(W)[h_1(W) - (\rho_d)\delta(r)] \\
&= h_1^2(W) - (\rho_d)\delta(r)h_1(W)
\end{aligned}$$

(7.38)

The 1st term is the effective or resultant PSF for perfect alignment of both objectives of linear apertures, while the 2nd term $= (\rho_1)\delta(r) \, h_1(W)$ is the aberration due to misalignment corresponding to the 2nd objective.

Following a similar procedure for the quadratic aperture, we obtain the following result for the effective PSF:

$$h_4(W; \rho_d) = h_3(W) - 2(\rho_d)h_1(W) + \rho_d^2 \delta(r)$$

$$\begin{aligned}
h_{\text{eff}}(\text{quadratic apert.}) &= h_3(W) \cdot h_4(W; \rho_d) \\
&= h_3(W)[h_3(W) - 2(\rho_d)h_1(W) + \rho_d^2(r)] \\
&= h_3^2(W) - 2(\rho_d)h_1(W)h_3(W) + \rho_d^2(r)h_3(W)
\end{aligned}$$

(7.39)

The 1st term in Eq. (7.39) represents the effective or resultant PSF for perfect alignment of both objectives of quadratic apertures, while the sum of the 2nd and 3rd terms is the aberration due to misalignment of the 2nd objective.

Hence, the misalignment aberration term due to the shifted quadratic aperture is given as follows:

$$g(W; \rho_d) = \left[\rho_d^2 \delta(r) - 2(\rho_d) h_1(W)\right] h_3(W) \tag{7.40}$$

For a perfectly aligned optical system for the objectives of the CSLM, the misaligned term in Eq. (7.40) has disappeared since $\rho_d = 0$, as expected.

Following the analysis in Ref. [8] for circular apertures where the second objective of the CSLM is subjected to tilting and a lateral shift, we obtain the following results for the misaligned linear and quadratic apertures:

For a collimated laser beam, if the 2nd objective is subjected to only a lateral shift ρ_d, the radial Bessel integral is written for the linear amplitude aperture as follows:

$$h_2(W; \rho_d) = 2\pi \int\limits_{-\rho_{02}+\rho_d}^{\rho_{02}+\rho_d} \rho J_0\left(\frac{2\pi\rho r}{\lambda f}\right) \rho d\rho \tag{7.41}$$

Using the properties of the Bessel function and integration by parts, we finally obtain the following result for a shifted objective with a linear aperture:

$$h_{2\,\text{linear}}(W; \rho_d) = 4(\rho_{02} + \rho_d)^3 \left\{ \left[\frac{J_1(W_H)}{W_H} + \frac{J_0(W_H)}{W_H^2} - 2\sum_i \frac{J_i(W_H)}{W_H^3}\right] \right.$$
$$\left. - \left(\frac{\varepsilon - 1}{\varepsilon + 1}\right)^3 \left[\frac{J_1(W_L)}{W_L} + \frac{J_0(W_L)}{W_L^2} - 2\sum_i \frac{J_i(W_L)}{W_L^3}\right] \right\} \tag{7.42}$$

For the shifted quadratic aperture, we solve the following radial integral:

$$h_4(W; \rho_d) = 2\pi \int\limits_{-\rho_{02}+\rho_d}^{\rho_{02}+\rho_d} \rho^2 J_0\left(\frac{2\pi\rho r}{\lambda f}\right) \rho d\rho \tag{7.43}$$

We finally obtain the PSF for the shifted objective, which has a quadratic aperture, as follows:

$$h_{4\,\text{quadratic}}(W; \rho_d) = 4\pi(\rho_{02} + \rho_d)^4 \left\{ \left[\frac{J_1(W_H)}{W_H} - 2\frac{J_2(W_H)}{W_H^2}\right] \right.$$
$$\left. - \left(\frac{\varepsilon - 1}{\varepsilon + 1}\right)^4 \left[\frac{J_1(W_L)}{W_L} - 2\frac{J_2(W_L)}{W_L^2}\right] \right\} \tag{7.44}$$

The PSF results for the shifted objective provided with linear and quadratic apertures in Eqs. (7.24, 7.26) are compared with the shifted objective corresponding to the circular aperture represented by Eq. (7.45):

$$h_{2\,\text{circular}}(W; \rho_d) = 4\pi(\rho_{02} + \rho_d)^2 \left\{ \left[\frac{J_1(W_H)}{W_H} \right] - \left(\frac{\varepsilon - 1}{\varepsilon + 1} \right)^2 \left[\frac{J_1(W_L)}{W_L} \right] \right\} \quad (7.45)$$

where $\varepsilon = \frac{\rho_d}{\rho_{02}} \ll 1$ in Eqs. (7.25), (7.27), and (7.45).

The upper and lower reduced coordinates are written as follows:

$$W_H = \frac{2(\rho_{02} + \rho_d)}{\lambda f} r = \frac{2}{\lambda} \text{N.A.} \left(1 + \frac{\rho_d}{\rho_{02}} \right) r \quad (7.46)$$

$$W_L = \frac{2\pi(-\rho_{02} + \rho_d)}{\lambda f} r = \frac{2}{\lambda} \text{N.A.} \left(\frac{\rho_d}{\rho_{02}} - 1 \right) r \quad (7.47)$$

The numerical aperture corresponding to the second objective is, $\text{NA} = \frac{\rho_{02}}{f}$, where f is the focal length.

The CTF for the CSLM is computed from the convolution product corresponding to the apertures of the microscope objectives. This gives a maximum value of unity in the center, which decreases until it reaches zero when the two apertures are completely separated. When the microscope objectives are affected by a displacement in the plane, the maximum value of the CTF decreases, as expected. An empirical formula for the maximum value of the CTF is given corresponding to one aligned objective with a quadratic or linear aperture, while the second has the same aperture distribution but is displaced by $\delta\rho$. The aperture radius $= \rho_0$

$$C(\rho) = P_1(\rho) * P_2(\nu) \quad (7.48)$$

Then, $C_{\text{max}}(= 0) = 1$; two aligned objectives, each having either a linear or quadratic aperture.

In the case of one aligned objective of one aligned linear or quadratic aperture and the 2nd objective having a misaligned linear or quadratic aperture, the maximum value of the CTF is expressed by the following empirical formula:

$$C_{\text{max}}(\delta) = \exp[-\alpha\pi(\delta\rho/\rho_0)] \quad (7.49)$$

where α is a parameter, whose value is dependent on the aperture distribution. The maximum value of the CTF becomes unity when the displacement $\delta\rho = 0$.

References

1. M. Minsky, U.S. Patent 3013467, Micros. Apparatus Dec. 19 (1961)
2. P. Davidovits, M.D. Egger, Scanning laser microscope. Nature **223**, 831 (1969)
3. G. Nomarski, J. Opt. Soc. Am. **65**(10), 1166 (1975)
4. C.J.R. Sheppard, A. Choudhary, Image formation in the scanning microscope. Opt. Acta **24**, 1051–1073 (1977)

5. C.J.R. Sheppard, The use of lenses with annular aperture in scanning optical microscopy. Optik **48**, 329–334 (1977)
6. J. Cox, C.J.R. Sheppard, T. Wilson, Improvement in resolution by nearly confocal microscopy. Appl. Opt. **21**, 778–781 (1982)
7. T. Wilson, J.N. Gannaway, Imaging properties and applications of scanning optical microscopes. Appl. Phys. **22**, 119–128 (1980)
8. C.J.R. Sheppard, T. Wilson, Imaging properties of annular lenses. Appl. Opt. **18**(22), 3764–3769 (1979)
9. C.J.R. Sheppard, T. Wilson, Fourier imaging of phase information in conventional and scanning microscopes. Phil. Trans. Roy. Soc. A **295**, 513–536 (1980)
10. A.M. Hamed, J.J. Clair, Image and super-resolution in optical coherent microscopes. Optik **64**, 277–284 (1983)
11. A.M. Hamed, J.J. Clair, Studies on optical properties of confocal scanning optical microscope using pupils with radially transmission distribution. Optik **65**, 209–218 (1983)
12. A.M. Hamed, Design of a cascaded black and linear distribution (CBLD) in circular aperture (application on confocal scanning laser microscope (CSLM). Int. Conf. in Physics, Houston USA, 28–29 September 2019
13. A.M. Hamed, Design of some heterogeneous apertures and computation of resolution. IJPOT **4**, 13–19 (2018)
14. AM. Hamed, Design of a Cascaded Black–Linear Distribution (CBLD) in circular aperture and its application on Confocal Laser Scanning Microscope (CSLM). Am. J. Optics Photonics **3** (2019)
15. A.M. Hamed, Improvement of point spread function (PSF) using linear-quadratic aperture. Optik **131**, 838–849 (2017). https://doi.org/10.1016/j.ijleo.2016.11.201
16. A.M. Hamed, A study on misaligned modulated apertures in Confocal Scanning Laser Microscope, accepted for oral presentation with good review comments after the review process, in *International Conference in Physics*, Houston, USA, 28–29 September 2019.https://helics group.net/conferences/addons/PhysicsUSA2019/29
17. A.M. Hamed, A hyper-resolving polynomial aperture and its application in microscopy. J. Basic Appl. Sci. **11** (2022)
18. A.M. Hamed, Modulated apertures and resolution in microscopy. Springer Briefs in Applied Sciences and Technology (2023). ISBN: 978-3-031-47552-8
19. A.M. Hamed, Aberration studies utilizing an optoelectronic coherent microscope. Optik **67**, 279–290 (1984)
20. A.M. Hamed, Exenteration errors combined with wavefront aberration in a coherent scanning microscope. Optik **82**, 1–4 (1989)
21. A.M. Hamed, T. Al-Saeed, A study on misaligned modulated apertures in confocal scanning laser microscope (CSLM). Int. J. Eng. Res. Tech. **7**(3), 9–27 (2019)

Chapter 8
Aberration Studies Using a Confocal Scanning Laser Microscope

8.1 Introduction

An optoelectronic coherent microscope (O.E.C.M.) referred to as confocal scanning laser microscopy (CSLM) in this chapter to study aberrations. The two microscope objectives are arranged in tandem, as shown in Fig. 8.1. This microscope (CSLM) uses a focused laser beam where the object found in the back focal plane of the first objective lens P_1 is mechanically scanned in the spatial plane defined by $z = $ constant. The second objective lens P_2 is used to image the object information. The optical image is transformed into an electronic image by employing a quadratic detector and hence displayed on a cathode ray oscilloscope (C.R.O.). Synchronization between the scanning beam of the C.R.O. and the mechanical scanning of the object is required to visualize the image over the oscilloscope screen. For coherent detection using the former microscope a pinhole is necessary to be found in the imaging plane with dimensions calculated easily from the diffraction pattern of the objective lens P_2. The resolution of this O.E.C.M. depends first on the optical resolution, which is computed from the resultant point spread function. $h_r = h_1 \cdot h_2$, and second, on the electronic resolution. This O.E.C.M. gives better resolution than scanning the light spot across a stationary object because the beam in the former case is always on- the axis, thus eliminating the off-axis aberration [1–3] of the objectives P_1 and P_2 and consequently ensuring that the optical resolution is invariant over the entire area of the image.

Many authors in [2, 3] applied the intensity criterium for studying truncated Gaussian apertures using the conventional optical microscope to calculate the image intensity degradation (defined as the square of the ratio of the diffracted amplitude in the presence of aberration to the diffracted amplitude for the non-aberrated system). In addition, it has been demonstrated that the optical resolution may be improved by using the microscope of type (2) [4–10] which we call O.E.C.M. Hamed, et al. [11–15] showed theoretically that further improvement of the resolution in the O.E.C.M. may be obtained by applying amplitude and phase modulations of the apertures.

© The Author(s), under exclusive license to Springer Nature Switzerland AG 2025
A. M. Hamed, *Studies on the Confocal Laser Microscope*,
SpringerBriefs in Applied Sciences and Technology,
https://doi.org/10.1007/978-3-031-87275-4_8

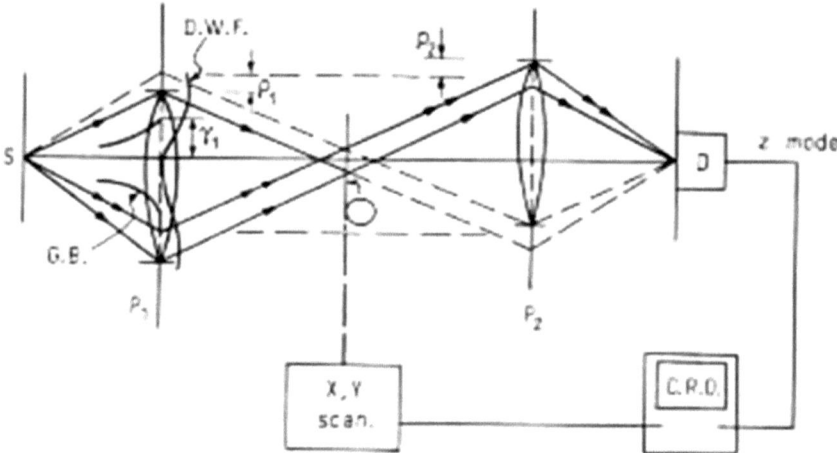

Fig. 8.1 Schematic diagram of the confocal scanning laser microscope provided with laterally displaced truncated Gaussian apertures. O: transparent object, P_1 and P_2 objective apertures, S: pinhole source, D: pinhole detector, G.B: Gaussian beam, γ: truncation parameter, D. W.F.: distortion wavefront

Recently, the contribution of the aperture modulation and resolution in microscopy is presented in [16]. Since the design of an optical system requires information about the diffraction intensity the Strehl ratio giving the image intensity degradation may be considered as a useful criterium for an image quality.

In this chapter, I studied the effect of the lateral and longitudinal displacements of the microscope objectives on the resultant diffraction pattern by applying the Strel ratio definition of the image intensity degradation using a confocal scanning laser microscope. I used two interesting pupils, the uniform circular aperture, and the truncated Gaussian pupil.

In the next section, I will present the theoretical analysis, followed by the theoretical results, and I will conclude the discussion.

Before I started the analysis, it was interesting to present the basic differences between the conventional microscope and the confocal microscope.

- The classical microscope is incoherently illuminated, while the confocal microscope is coherently illuminated using a pinhole source.
- The image detected by the classical microscope is coherently visualized by employing an objective lens and eyepiece. The detection in the confocal microscope is obtained coherently through a pinhole inserted in contact with the detector in the imaging plane.
- We cannot assume that an image can be visualized without scanning both the object and the electronic beam emitted from the cathode ray tube. For this reason, I have chosen the name of an optoelectronic coherent microscope (O.E.C.M.).

In general, in the O.E.C.M. Both the illumination and the detection are coherent, while in a classical optical microscope, only the detection is coherent, while the illumination of the object is incoherent. In addition, the object scans in the microscope under consideration while it is stationary in the classical microscope. The two microscope objectives contributed equally to the resolution of the confocal microscope or the O.E.C.M. However, only the objective imaging lens in the classical microscope is responsible for the resolution.

The image intensity for each of the four equivalent types is shown in Fig. 8.2 of the confocal optoelectronic scanning microscope and is represented as follows:

$$I = |O \otimes h_1 h_2|^2$$

8.2 Theoretical Analysis

8.2.1 Diffraction Intensity Using the CSLM Provided with Laterally Displaced Truncated Gaussian Apertures [14]

The normalized point spread function for a truncated Gaussian aperture laterally displaced by a distance ρ_1 from the optical axis is calculated as follows:

$$h_1 = \frac{\int_0^1 \int_0^{2\pi} \exp[-(\rho - \rho_1)^2 / \gamma_1^2] \exp[jk\phi_1(\rho)] \rho d\rho d\theta}{\int_0^1 \int_0^{2\pi} \exp\left[-\left(\frac{\rho}{\gamma_1}\right)^2\right] \rho d\rho d\theta} \tag{8.1}$$

In formula (8.1), we consider that: ρ_1 is the lateral aberration due to the misalignment of the objective lens P_1 to the optical axis. The displaced Gaussian amplitude is given by $\exp\left[-\frac{(\rho-\rho_1)^2}{\gamma_1^2}\right]$, where γ_1 is the truncation limit of the Gaussian function and ρ is the radial coordinate in the aperture plane. The wave distortion phase term introduced by the objective lens P_1 is given by $\phi_1(\rho)$.

The normalized Gaussian function in the denominator of Eq. (8.1) is solved to give the following:

$$\int_0^1 \int_0^{2\pi} \exp\left[-\left(\frac{\rho}{\gamma_1}\right)^2\right] \rho d\rho d\theta = 2\pi \int_0^1 \exp -\left[\left(\frac{\rho}{\gamma_1}\right)^2\right] \rho d\rho$$

$$= \pi \gamma_1^2 \left[1 - \exp\left(-\frac{1}{\gamma_1^2}\right)\right] = \pi N(\gamma_1) \tag{8.2}$$

By substituting Eq. (8.2) into Eq. (8.1), we write:

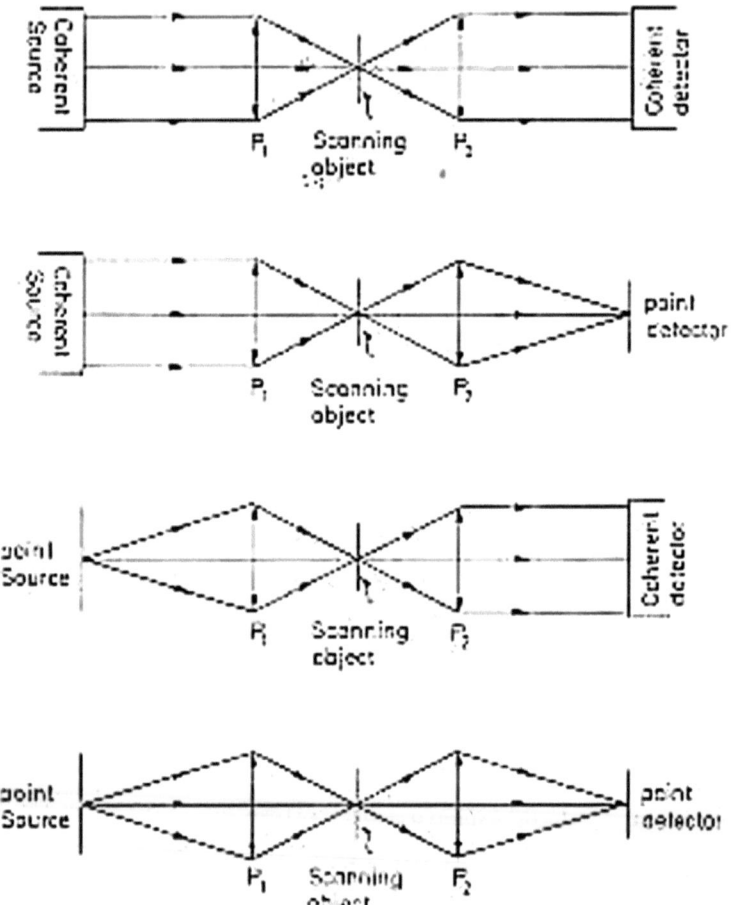

Fig. 8.2 Four equivalent types of confocal scanning laser microscopes with coherent illumination and detection

$$h_1 = \frac{\exp\left[-\left(\frac{\rho_1}{\gamma_1}\right)^2\right]}{\pi N(\gamma_1)} \int_0^1 \int_0^{2\pi} \exp\left[jk\phi_1(\rho) - \left(\frac{\rho}{\gamma_1}\right)^2 + \left(\frac{2\rho\rho_1}{\gamma_1^2}\right)\right]\rho\,d\rho\,d\theta \qquad (8.3)$$

Assuming that $2\rho\rho_1 \ll \gamma_1^2$, we can write Eq. (8.3) as follows:

$$h_1 = \frac{\exp\left[-\left(\frac{\rho_1}{\gamma_1}\right)^2\right]}{\pi N(\gamma_1)} \left\{ \int_0^1 \int_0^{2\pi} \rho \exp\left[jk\phi_1(\rho) - \left(\frac{\rho}{\gamma_1}\right)^2 \right] d\rho d\theta \right.$$

$$\left. + \frac{2\rho_1}{\gamma_1^2} \int_0^1 \int_0^{2\pi} \rho^2 \exp\left[jk\phi_1(\rho) - \left(\frac{\rho}{\gamma_1}\right)^2 \right] d\rho d\theta \right\} \qquad (8.4)$$

Equation (8.4) is obtained after developing the exponential term of $2\rho\rho_1/\gamma_1^2$, retaining only the first two terms. After a long calculation, one then obtains the following closed expression for the normalized PSF valid to a good approximation at the center of the diffraction focal plane, i.e., we finally obtain:

$$h_1 = \exp\left[-\left(\frac{\rho_1}{\gamma_1}\right)^2\right]\left\{\left[1 + jk\overline{\Delta_1} - \frac{k^2}{2}\overline{\Delta_1^2}\right]\right.$$

$$+ \frac{4\rho_1}{\gamma_1^2 N(\gamma_1)} \left[\sum_{m=0} \frac{(-1)^m}{m!(2m+3)} \left(\frac{1}{\gamma_1}\right)^{2m} \right.$$

$$+ jk\, a_n \sum_{m=0}^{\infty} \frac{(-1)^m}{m![2(n+m)+3]} \left(\frac{1}{\gamma_1}\right)^{2m}$$

$$\left.\left. - \frac{k^2}{2} a_n^2 \sum_{m=0} \frac{(-1)^m}{m![2(2n+m)+5]} \left(\frac{1}{\gamma_1}\right)^{2m} \right]\right\} \qquad (8.5)$$

The nth moment of the Gaussian aberration function $\Delta_1^n(\rho) = \phi_1^n(\rho)\exp-\left(\frac{\rho}{\gamma_1}\right)^2$ is defined as:

$$\overline{\Delta_1^n(\rho)} = \frac{1}{\pi N(\gamma_1)} \int_0^1 \int_0^{2\pi} \exp\left[-\left(\frac{\rho}{\gamma_1}\right)^2\right]\phi_1^n(\rho)\rho d\rho d\theta \qquad (8.6)$$

I obtained Eq. (8.5) after developing the exponential term of the wave aberration function retaining terms up to the second order, which is valid for the low numerical aperture (N.A.) of the objectives.

A similar expression can be obtained for the second objective lens if the lateral displacement of ρ_2 is in the opposite direction to the lateral displacement ρ_1 caused by the first objective. Consequently, the normalized diffracted amplitude gives:

$$
\begin{aligned}
h_2 = {} & \exp\left[-\left(\frac{\rho_2}{\gamma_2}\right)^2\right]\left\{\left[1+jk\overline{\Delta_2}-\frac{k^2}{2}\,\overline{\Delta_2^2}\right]\right. \\
& -\frac{4\rho_2}{\gamma_2^2 N(\gamma_2)}\left[\sum_{m=0}^{\infty}\frac{(-1)^m}{m!(2m+3)}\left(\frac{1}{\gamma_2}\right)^{2m}\right. \\
& \hspace{3cm} \left. +jk\,a_n\sum_{m=0}^{\infty}\frac{(-1)^m}{m![2(n+m)+3]}\left(\frac{1}{\gamma_2}\right)^{2m} \right. \\
& \hspace{3cm} \left.\left.-\frac{k^2}{2}\,a_n^2\sum_{m=0}^{\infty}\frac{(-1)^m}{m![2(2n+m)+5]}\left(\frac{1}{\gamma_2}\right)^{2m}\right]\right\}
\end{aligned}
\tag{8.7}
$$

Equations (8.5, 8.7) are obtained for a single aberration, i.e., $\phi^n(\rho)=a_n\rho^{2n}$, n is an even number. $N(\gamma_2)$ and $\overline{\Delta_2^n}$ are defined as before, where suffix 1 is replaced by suffix 2.

The resultant diffraction intensity for two symmetric laterally displaced apertures is calculated from $I=|h_1\cdot h_2|^2$, and we obtained the following formula:

$$
\begin{aligned}
I(a_n; \rho_1) = {} & \exp\left[-4\left(\frac{\rho_1}{\gamma_1}\right)^2\right]\left\{\left[1-k^2\left(\overline{\Delta_1^2}-(\overline{\Delta_1})^2\right)-\left(\frac{4\rho_1}{\gamma_1^2 N(\gamma_1)}\right)^2\right.\right. \\
& \left(\left[\sum_{m=0}^{\infty}\frac{(-1)^m}{m!(2m+3)}\left(\frac{1}{\gamma_1}\right)^{2m}-\frac{k^2}{2}\,a_n^2\sum_{m=0}^{\infty}\frac{(-1)^m}{m![2(2n+m)+5]}\left(\frac{1}{\gamma_1}\right)^{2m}\right]^2\right. \\
& \left.\left.-k^2\,a_n^2\left[\sum_{m=0}^{\infty}\frac{(-1)^m}{m![2(n+m)+3]}\left(\frac{1}{\gamma_1}\right)^{2m}\right]^2\right]\right)+4\,k^2\,a_n^2\left(\frac{4\rho_1}{\gamma_1^2 N(\gamma_1)}\right)^4 \\
& \left[\sum_{m=0}^{\infty}\frac{(-1)^m}{m!(2m+3)}\left(\frac{1}{\gamma_1}\right)^{2m}-\frac{k^2}{2}\,a_n^2\sum_{m=0}^{\infty}\frac{(-1)^m}{m![2(2n+m)+5]}\left(\frac{1}{\gamma_1}\right)^{2m}\right]^2 \\
& \left.\times\left[\sum_{m=0}^{\infty}\frac{(-1)^m}{m![2(n+m)+3]}\left(\frac{1}{\gamma_1}\right)^{2m}\right]^2\right\}
\end{aligned}
\tag{8.8}
$$

For perfect alignment of the optical components, i.e., $\rho_1=\rho_2=0$ the diffraction intensity is localized at the geometrical focus and Eq. (8.8) reduces to

$$
I(a_n)=\left[1-k^2\left(\overline{\Delta_1^2}-(\overline{\Delta_1})^2\right)\right]^2
\tag{8.9}
$$

Physically Eq. (8.9) is expected since the resultant diffraction intensity is calculated by.

$I=|h_2\cdot h_2|^2$ which is the case of confocal operation of the O.E.C.M. and consequently the result given by Eq. (8.9) is the square of the Lowenthal [2] result which is valid for the classical optical systems.

8.2.1.1 Marechal Results for the Conventional Microscope

Marechal [17] obtained this result for image degradation using a uniform circular aperture:

$$i_M = 1 - k^2\left[\overline{\phi^2} - \left(\overline{\phi}\right)^2\right] \tag{8.10}$$

Equation (8.10) can be easily obtained using the Strel definition of image intensity degradation, where the n-th moment of $\phi(\rho)$ is given by:

$$\overline{\phi}_n = \frac{1}{\pi} \int_0^1 \int_0^{2\pi} \phi_n(\rho)\rho d\rho d\theta \tag{8.11}$$

Lowenthal [2] obtained the following result, which is valid for conventional optical systems with Gaussian apertures.

$$i_L = 1 - k^2\left[\overline{\Delta^2} - \left(\overline{\Delta}\right)^2\right] \tag{8.12}$$

8.2.1.2 Author's Results Using the CSLM Provided with a Uniform Circular Aperture

Assuming that the amplitude transmission function from the aperture is unity, the intensity degradation can be determined as follows:

$$i_A = \left|\frac{\int_0^1 \int_0^{2\pi} A_1(\rho)\exp(jk\phi_1(\rho))\rho d\rho d\theta \int_0^1 \int_0^{2\pi} A_2(\rho)\exp(jk\phi_2(\rho))\rho d\rho d\theta}{\int_0^1 \int_0^{2\pi} A_1(\rho)\rho d\rho d\theta \int_0^1 \int_0^{2\pi} A_2(\rho)\rho d\rho d\theta}\right|^2 \tag{8.13}$$

where $A_2(\rho) = A_1(\rho) = 1$.

We believe that retaining the terms of $\phi(\rho)$ up to the 3rd order during the development of its exponential will give better results for moderate apertures of the objectives. We pay attention that the precedent results [2, 15] were obtained keeping only terms of $\phi(\rho)$ up to the 2nd order which is satisfied only for low N.A. In the former case, we obtain this result:

$$i_{author} = 1 - k^2\left\{\left[\overline{\phi_1^2} - \left(\overline{\phi_1}\right)^2\right] + \left[\overline{\phi_2^2} - \left(\overline{\phi_2}\right)^2\right]\right\} + k^4\left\{\left[\overline{\phi_1^2} - \left(\overline{\phi_1}\right)^2\right]\left[\overline{\phi_2^2} - \left(\overline{\phi_2}\right)^2\right]\right\} \tag{8.14}$$

For equal aberration of the two microscope objectives, i.e., $\phi_1(\rho) = \phi_2(\rho)$. Then, we obtain from Eq. (8.14) this formula:

$$i = \left\{ 1 - k^2 \left[\overline{\phi^2} - \left(\overline{\phi} \right)^2 \right] \right\}^2 \tag{8.15}$$

For a wavefront distortion $\phi(\rho)$ described by an even power series in ρ; i.e., $\phi(\rho)$ gives all orders of spherical aberration as:

$$\phi(\rho) = \sum_{n=2}^{N} a_n \rho^{2n} \tag{8.16}$$

In this case, by substituting Eq. (8.16) with Eq. (8.15), we obtain the following equation.

$$i = \left\{ 1 - k^2 \left[\sum_{m=0}^{M} \frac{b_m}{(m+1)} - \left(\sum_{n=0}^{n} \frac{a_n}{(n+1)} \right)^2 \right] \right\}^2 \quad m = 2n, \ b_m = a_n^2 \tag{8.17}$$

For a single aberration, we obtain this result:

$$i = \left(1 - \left(\frac{2\pi}{\lambda} \right)^2 a_n^2 \left\{ \frac{1}{2n+1} - \frac{1}{(n+1)^2} \right\} \right)^2 \tag{8.18}$$

8.3   Results and Discussion

The resultant diffraction intensity is calculated for laterally displaced objectives using an optoelectronic coherent microscope as shown in Fig. 8.1. We have assumed Gaussian apertures for the microscope objectives with truncation parameter γ. The amount of lateral displacement is $\rho_1 = 100$ m. The intensity values are computed for different values of a_n (spherical aberration coefficient) using the analytical formula (8.8) and graphically represented as in Fig. 8.3 where a set of four curves is drawn with the following parameters: $\rho_1 = 100$ m, $\gamma = 1, 2, 4$, and 8 cm. We can see, referring to Fig. 8.3 that the intensity degrades little as the truncation width γ_1 of the Gaussian function increases which is physically acceptable since it depends on the numerical aperture calculated by $NA = \frac{\rho_0}{f} \exp - \left(\frac{\rho}{\gamma} \right)^2$ for truncated Gaussian function.

In the limiting case, when $\gamma \to \infty$ gives the results corresponding to uniform circular apertures as shown in Fig. 8.4 where two curves are drawn one for the O.E.C.M. and the other for the conventional microscope assuming that $\rho_1 = 0$ for the confocal arrangement.

Fig. 8.3 On-axis intensity degradation for laterally displaced Gaussian apertures versus the spherical aberration coefficient a_n with truncation width $\gamma_1 = 1, 2, 4,$ and 8 using a confocal scanning laser microscope

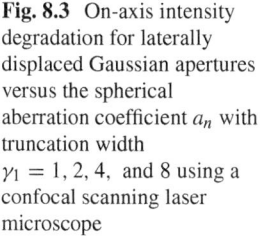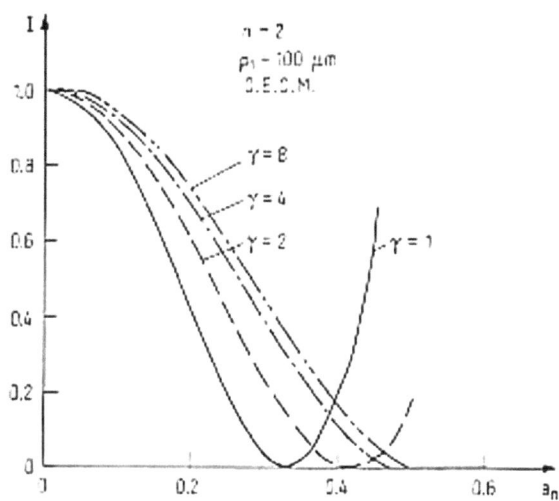

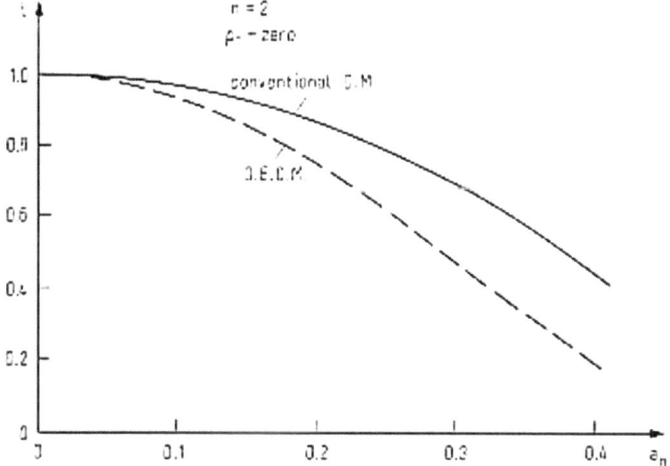

Fig. 8.4 On-axis intensity degradation for a uniform circular aperture versus the spherical aberration coefficient a_n. The continuous curve is given for the conventional optical microscope, and the discontinuous curve is given for the confocal scanning laser microscope ($\rho_1 = 0$)

8.4 Conclusion

Referring to Fig. 8.4, we can conclude that for a certain value of a_n or the spherical aberration coefficient, the diffracted intensity degrades much for the CSLM provided with either a uniform circular aperture or Gaussian pupils compared to the corresponding results for conventional optical systems. These obtained results are expected since the fractional normalized diffraction intensity is calculated with the help of Strel's definition by squaring the diffraction results obtained for the C.O.S.

In addition, we conclude that the mean square deviation of the wavefront from the reference sphere $\Delta\phi$ is less than $\lambda/5$ for a degradation of 20%. The CSLM was compared to $\Delta\phi < \lambda/14$ for the same intensity degradation using the classical optical system.

The most interesting property of this confocal microscope is the resolution improvement, as confirmed by Sheppard et al., and Hamed et al., while the image contrast is much lower than that obtained by the C.O.S.

References

1. C.J.R. Sheppard, Scanning optical microscope. Electronics & Power **26**, 166–172 (1980)
2. D.D. Lowenthal, Marechal intensity criteria modified for Gaussian beams. Appl. Opt. **13**, 2126 (1974)
3. J.W. Goodman, *Introduction to Fourier Optics* (McGraw-Hill Book Comp, New York, 1968)
4. C.J.R. Sheppard, A. Choudhary, Opt. Acta **24**, 1051–1059 (1977)
5. J.J. Cox, C.J.R. Sheppard, T. Wilson, Improvement in resolution by confocal microscopy. Appl. Opt. **21**, 778–781 (1982)
6. C.J.R. Sheppard, X.Q. Mao, Confocal microscopes with slit apertures. J. Mod. Opt. **35**, 1169–1185 (1988)
7. C.J.R. Sheppard, Supper resolution in confocal imaging. Optik **80**, 53–54 (1988)
8. C.J.R. Sheppard, M. Gu, Improvement of axial resolution in confocal microscopy using annular pupil. Opt. Commun. **84**, 7–13 (1991)
9. M. Gu et al., Optimization of axial resolution in confocal imaging using annular pupils. Optik **93**, 87–90 (1993)
10. G. Cox, C.J.R. Sheppard, Practical limits of resolution in confocal and nonlinear microscopy. Microsc. Res. Tech. **63**, 18–22 (2004)
11. J.J. Clair, A.M. Hamed, Theoretical studies on optical coherent microscope. Optik **64**, 133–141 (1983)
12. A.M. Hamed, J.J. Clair, Image and supper resolution in optical coherent microscopes. Optik **64**, 272–284 (1983)
13. A.M. Hamed, J.J. Clair, Studies on optical properties of confocal scanning optical microscope using pupils with radially transmission distribution. Optik **65**, 209–218 (1983)
14. A.M. Hamed, Aberration studies utilizing an optoelectronic coherent microscope. Optik **67**, 279–290 (1984)
15. A.M. Hamed, Exenteration errors combined with wavefront aberration in a coherent scanning microscope. Optik **82**, 1–4 (1989)
16. A.M. Hamed, Modulated Apertures and Resolution in Microscopy. Springer Briefs in Applied Sciences and Technology. ISBN 978-3-031-47552-8 (2023)
17. A. Marechal, Study of the combined effects of diffraction and geometrical aberrations on the image of a luminous point. Rev. Opt. **26**, 257 (1947)

Chapter 9
Spatial Coherence in Confocal Microscopy for Quadratically Radially Distributed Apertures

We study the spatial coherence problem using an amplitude modulation applied to confocal imaging systems. This type of modulation assumes a quadratic radial distribution. The mutual coherence intensity or the coherence factor is calculated and compared with the results obtained for clear circular apertures.

9.1 Introduction

When confocal imaging systems are used, many problems arise. Among these is the problem of spatial coherence. This problem has been treated by many authors, and it has been proven that the complex degree of spatial coherence is obtained by applying the Fourier transform (FT) to the intensity distribution of the extended source. The theory of partial coherence has been applied to ordinary microscopes characterized by a single objective lens that is responsible for the resolution.

However, in the former case of confocal systems, the condenser lens is replaced by an aberration-free objective lens. Hence, the resolution is determined from the resultant point spread function (P.S.F), namely, $h_r = h_1 \cdot h_2$, where h_1 is the first objective and h_2 is the second objective. In this confocal arrangement, the scanned object is located at the common short focus of the aberration-free objective L_1, L_2 and is arranged in tandem. The image of a point object is obtained from the modulus square of the convolution product of the Dirac delta function and the resultant point spread function of the imaging system [1–6].

This chapter studies amplitude modulation applied to confocal systems for the following reasons. Perfect coherent detection is considered an ideal case, and the aperture of the detector has a physically finite size. Hence, the coherent detection process is very difficult to realize. It seems that partially coherent detection is very suitable in this case. Second, this study of spatial coherence using amplitude modulation or quadratic radial distribution is applied to confocal systems. Hence, a truncated

© The Author(s), under exclusive license to Springer Nature Switzerland AG 2025   105
A. M. Hamed, *Studies on the Confocal Laser Microscope*,
SpringerBriefs in Applied Sciences and Technology,
https://doi.org/10.1007/978-3-031-87275-4_9

quadratic radial function is considered. The theoretical analysis is followed by the results and discussion.

9.2 Analysis

In this chapter, we calculate the mutual coherence intensity.

In this case, a clear circular aperture is assumed for the pupil P_1, and a truncated quadratic radially distributed function is assumed for the second pupil P_2 in the confocal arrangement, as shown in Fig. 9.1.

It was previously shown that the image of a point is calculated as [7, 8]:

$$I(w) = |\, g(w) \otimes h_r(w)|^2 \qquad (9.1)$$

where $h_r = h_1 \cdot h_2$ is the resultant PSF of the imaging system, and $\otimes$ is a symbol for the convolution product.

This ideal case of confocal imaging assumes that both the illumination and the detection are completely coherent. Coherent illumination is ensured by utilizing the laser source, while the coherence of the detector is not completely ensured, since real detectors are physically finite in size. Hence, this study on spatial coherence considers partially coherent detection.

It is well known that the PSF of a clear circular aperture is an Airy disk obtained by operating the FT upon P_1 defined as

$$P_1(\rho) = 1, \quad \text{when} \left| \frac{\rho}{\rho_{01}} \right| \le 1$$
$$= 0; \text{ otherwise} \qquad (9.2)$$

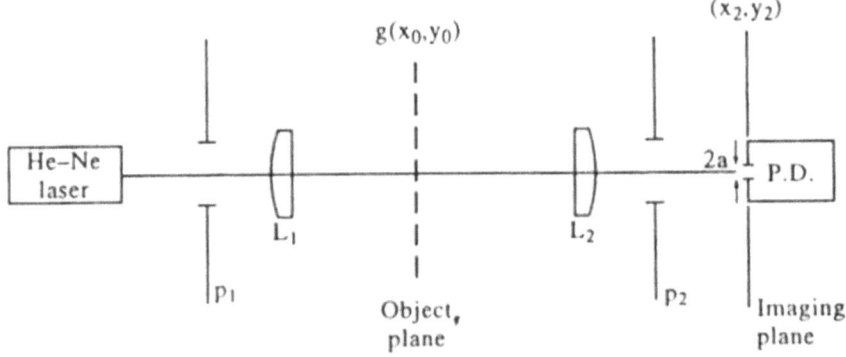

Fig. 9.1 Setup of the confocal optical system with a circular aperture for L_1 and a quadratic aperture for L_2. The diameter of the detector is 2a

where ρ_{01} is the radius of the aperture; hence, the PSF is obtained as follows:

$$h_1(r) = \text{F.T.}\{P_1(\rho)\} = \frac{2J_1(\alpha r)}{\alpha r} \tag{9.3}$$

where $\alpha = \left(\frac{2\pi}{\lambda}\right)$ NA, and NA is the numerical aperture.

Assuming a truncated radial distribution for the second aperture P_2, that is:

$$P_2(\rho) = \rho^2, \text{ when} \left|\frac{\rho}{\rho_{02}}\right| \leq 1 \tag{9.4}$$
$$= 0; \text{ otherwise}$$

where ρ_{02} is the radius of the second aperture.

The PSF is obtained by operating the FT upon P_2 as follows:

$$h_2(r) = \text{F.T.}\{P_2(\rho)\}$$
$$= 2 \int_0^{\rho_{02}} \rho^3 J_0(k\rho r)\,d\rho \tag{9.5}$$

Using integration by parts and recurrence relations, we finally obtain [8]:

$$h_2(r) = const.\left[\frac{J_1(w)}{w} - 2\frac{J_2(w)}{w^2}\right] \tag{9.6}$$

where $w = k\rho_{02}r$, and $\text{k} =$ is the propagation constant.

In the imaging plane. The complex amplitude for a stationary object $g(x_0, y_0)$ is calculated [9] as:

$$A_{\text{st.}}(x_2, y_2) = \iint_{-\infty}^{\infty} [h_1(x_0, y_0)\,g(x_0, y_0)]\,h_2(x_0 + x_2, y_0 + y_2)\,dx_0\,dy_0 \tag{9.7}$$

For a scanned object, the complex amplitude formed in the imaging plane is:

$$A_{\text{scan}}(x_2, y_2) = \left[h_1(x_s, y_s)h_2(x_s + x_2, y_s + y_2)\right] \otimes g(x_s, y_s) \tag{9.8}$$

where (x_s, y_s) are the Cartesian scanning coordinates.

The detected intensity distribution for the scanned object is obtained by taking the modulus square of the complex amplitude and then integrating it over the surface area of the detector, that is:

$$I(x_s, y_s; a) = \int\int_{-\infty}^{\infty} D(x_2, y_2)|A_{\text{scan}}(x_2, y_2)|^2 dx_2 dy_2 \tag{9.9}$$

where $D(x_2, y_2)$ is the sensitivity of the photodetector, and a circular aperture of radius a is assumed. In polar coordinates

$$D(x_2, y_2) = D(r_2) = 1 \text{ when } |\tfrac{r_2}{a}| \le 1$$
$$= \text{zero, otherwise}$$

(9.10)

Assuming that the object is composed of two luminescent points separated by a distance d, we can represent $g(x, y)$ as follows:

$$g(x, y) = \delta\left(x - \frac{d}{2}, y\right) + \delta\left(x + \frac{d}{2}, y\right)$$

(9.11)

Substituting Eqs. (9.10) and (9.11) into Eq. (9.9), we obtain the following result for the imaging intensity considering partially coherent detection:

$$I(r_s, ; a) = \int\limits_{-\infty}^{\infty}\int D(r_2)\Big[h_1\left(r_s - \frac{d}{2}\right) \cdot h_2\left(r_s + r_2 - \frac{d}{2}\right) + h_1\left(r_s + \frac{d}{2}\right)$$
$$\cdot h_2\left(r_s + r_2 + \frac{d}{2}\right)\Big]^2 r_2 dr_2$$

(9.12)

Equation (9.12) can be rewritten as follows:

$$I(r_s, ; a) = h_1^2\left(r_s - \frac{d}{2}\right)\int\limits_0^a h_2^2\left(r_s + r_2 - \frac{d}{2}\right) r_2 dr_2$$
$$+ h_1^2\left(r_s + \frac{d}{2}\right)\int\limits_0^a h_2^2\left(r_s + r_2 + \frac{d}{2}\right) r_2 dr_2$$
$$+ 2h_1\left(r_s - \frac{d}{2}\right)h_1\left(r_s + \frac{d}{2}\right)\gamma_{12}$$

(9.13)

where γ_{12} is the mutual coherence function and is computed as follows:

$$\gamma_{12} = \int\limits_0^a h_2\left(r_s + r_2 - \frac{d}{2}\right) \cdot h_2^*\left(r_s + r_2 + \frac{d}{2}\right) r_2 dr_2$$

(9.14)

For a stationary object, $r_s = 0$, γ_{12} becomes:

$$\gamma_{12} = \int\limits_0^a h_2\left(r_2 - \frac{d}{2}\right) \cdot h_2^*\left(r_2 + \frac{d}{2}\right) r_2 dr_2$$

(9.15)

By substituting Eq. (9.6) into Eq. (9.15), we finally obtain:

$$\pi_{12} = \int\limits_0^a \frac{J_1[\alpha(r_2 - \frac{d}{2})]}{\alpha(r_2 - \frac{d}{2})} \frac{J_1(r_2 + \frac{d}{2})}{\alpha(r_2 + \frac{d}{2})} r_2 dr_2$$

$$+4\int_0^a \frac{J_2\left[\alpha\left(r_2-\frac{d}{2}\right)\right]}{\left[\alpha\left(r_2-\frac{d}{2}\right)\right]^2}\frac{J_2\left(r_2+\frac{d}{2}\right)}{\left[\alpha\left(r_2+\frac{d}{2}\right)\right]^2}r_2 dr_2$$

$$-2\int_0^a \frac{J_1\left[\alpha\left(r_2-\frac{d}{2}\right)\right]}{\left[\alpha\left(r_2-\frac{d}{2}\right)\right]}\frac{J_2\left(r_2+\frac{d}{2}\right)}{\left[\alpha\left(r_2+\frac{d}{2}\right)\right]^2}r_2 dr_2$$

$$-2\int_0^a \frac{J_2\left[\alpha\left(r_2-\frac{d}{2}\right)\right]}{\left[\alpha\left(r_2-\frac{d}{2}\right)\right]^2}\frac{J_1\left(r_2+\frac{d}{2}\right)}{\left[\alpha\left(r_2+\frac{d}{2}\right)\right]}r_2 dr_2 \tag{9.16}$$

9.2.1 Special Case

For a clear circular aperture, γ_{12} in Eq. (9.16) becomes:

$$\gamma_{12}=\int_0^a \frac{J_1\left[\alpha\left(r_2-\frac{d}{2}\right)\right]}{\left[\alpha\left(r_2-\frac{d}{2}\right)\right]}\frac{J_1\left(r_2+\frac{d}{2}\right)}{\left[\alpha\left(r_2+\frac{d}{2}\right)\right]}r_2 dr_2 \tag{9.17}$$

9.3 Results and Discussion

The integral expressions obtained for the mutual coherence function or the coherence factor Eqs. (9.15) and (9.16) are numerically computed. A computer program is constructed to facilitate the computation of the above integrals. A graph of the coherence factor versus the two-point object (d) is plotted in Fig. 9.2. A set of four curves is drawn for a $= 0.3, 0.5, 0.7$, and 0.9 mm using a quadratic radially distributed aperture for the detector. Another set of three curves is plotted for a clear circular aperture, as shown in Fig. 9.3. The two curves shown in Fig. 9.4 are drawn for both the quadratic and the clear circular apertures for a detector aperture $a = 0.9$ mm.

The obtained results of the quadratic radial distribution are shown to vary slightly for $a \le 0.8$ mm, while a sharp change occurs in the coherence factor for $a \ge 0.9$ mm, as shown in Fig. 9.2. In the case of a clear circular aperture, a slight variation also occurs in the coherence factor for $a \le 0.9$ mm, while a sharp drop occurs for $a = 1.0$ mm, as shown in Fig. 9.3.

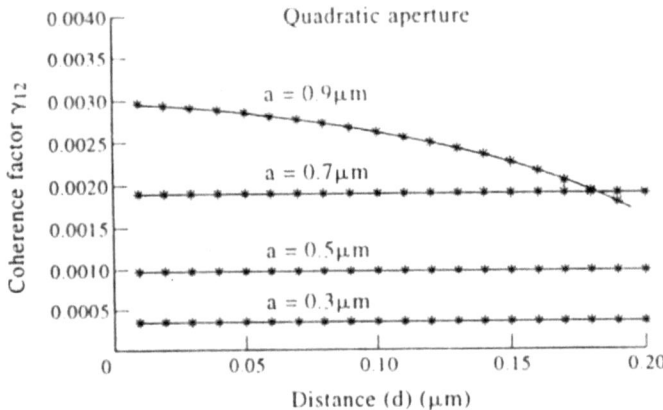

Fig. 9.2 Coherence factor versus distance (*d*) between the two-point object for different values of detector aperture (*a*) using quadratic aperture

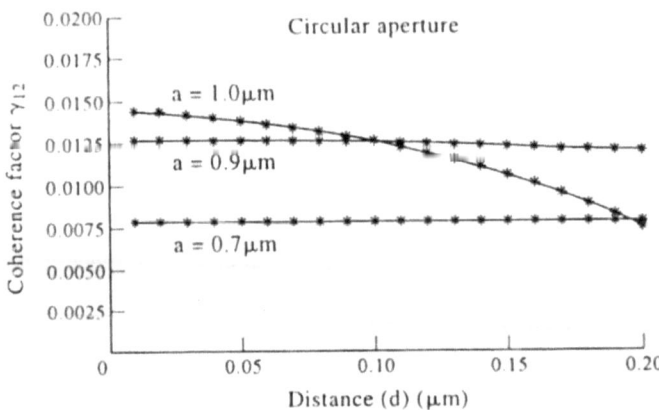

Fig. 9.3 Coherence factor versus distance (*d*) between two points for different values of the detector aperture (*a*) using a circular aperture

9.4 Conclusions

The optimal value of the coherence factor is attained for a value of $a < 0.9$ in this case of quadratic radial obstruction of the detector aperture. It is shown, from the obtained results, that the coherence factor decreases as the distance between the two-point objects increases in the two cases of quadratic modulation, and without modulation, as expected.

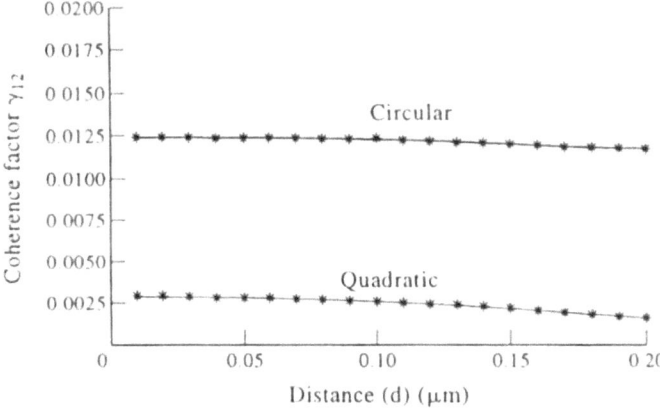

Fig. 9.4 Coherence factor versus distance (d) between two points for two apertures of different shapes ($a = 0.9$ mm)

References

1. H.H. Hopkins, P.M. Barham, The influence of the condenser on microscopic resolution. Proc. Phys. Soc. B **63**, 737 (1950)
2. H.H. Hopkins ll, The concept of partial coherence in optics. Proc. R. Soc. **A208**, 263 (1951)
3. M. Francon, S. Slansky, Coherence in Optics (1963)
4. G. Nomarski, M. Rousseau, Differential interference contrast J. Phys. Radium **16**, 13 or 9S (1955)
5. C.J.R. Sheppard, T. Wilson, Image formation in scanning microscopes with partially coherent source and detector. Opt. Acta **25**(4), 315–325 (1978)
6. C.J.R. Sheppard, T. Wilson, Multiple traversing of the object in the scanning microscope. Opt. Acta **27**(1980), 611–624 (1980)
7. C.J.R. Sheppard, T. Wilson, Image formation in confocal scanning microscopes. Optik **55**, 331–342 (1980)
8. A.M. Hamed, J.J. Clair, Studies on optical properties of confocal scanning optical microscope using pupils with radial transmission distribution. Optik **65**(3), 209–218 (1983)
9. A.M. Hamed, Optimization of spatial coherence in confocal optical systems. Opt. Laser Technol. **22**(2), 137–139 (1990)

Chapter 10
Spatial Coherence in Confocal Microscopy for Black and White Annular Apertures

We considered a black and white (B/W) annular aperture for the 2nd objective in the confocal microscope. For practical reasons, the coherent point detector in confocal imaging systems is replaced by a detector of finite size. We computed the spatial coherence factor for the B/W annulus using MATLAB code.

We obtained an empirical formula that represents the coherence factor versus the distance between two-point objects. We compared our results on spatial coherence in the case of B/W annulus with the previous results obtained in the case of circular and quadratic apertures using MATLAB.

10.1 Introduction

Early on, partial coherence theory was applied to ordinary optical microscopes characterized by a single objective lens that is responsible for the resolution, while the condenser lens is responsible for the illumination.

The spatial coherence problem was discussed earlier by many authors [1–5]. It was shown that the complex degree of spatial coherence is obtained by applying the Fourier transform to the intensity distribution of the extended sources.

The mutual coherence intensity is computed for confocal imaging systems by applying coherence theory to circular apertures for both microscope objectives in [6] and then computed considering the quadratic aperture [7, 8], where the quadratic aperture replaces the circular aperture for the 2nd objective lens.

In the former case of confocal imaging systems, it was shown that both objective lenses are responsible for the resolution since they are arranged in tandem where the scanned object is placed in the common short focus. Hence, the resultant point spread function (RPSF) is computed from the product of the PSF corresponding to each objective [9–12]. The effect of temporal and spatial coherence on resolution in full-field optical coherence tomography was described in [13]. The effect of the

© The Author(s), under exclusive license to Springer Nature Switzerland AG 2025   113
A. M. Hamed, *Studies on the Confocal Laser Microscope*,
SpringerBriefs in Applied Sciences and Technology,
https://doi.org/10.1007/978-3-031-87275-4_10

detector size on the PSF under a confocal microscope is shown. The 4Pi confocal point spread functions are shown for constructive and destructive interference of the collected wavefronts and are compared with the point spread functions of comparable confocal microscopes [14].

The degree of temporal coherence is necessary to obtain the monochromatic beam of the laser, which has a very small spectral width. Hence, the coherence time and coherence length are inversely proportional to the spectral width. For example, we obtain coherence length = 50 m for the He–Ne laser at spectral width = 1 MHz Spatial coherence is defined as the region of space for which the absolute difference in the optical field is less than π. Hence, in confocal laser scanning microscopy (CLSM), the illumination is coherent, while the detection is partially coherent to obtain sufficient detected light instead of point detection [7].

Both types act as three-dimensional imaging tools. The first is low temporal coherence microscopy and macroscopy, also known as optical coherence tomography (OCT), which is being used for medical diagnostics, particularly in ophthalmology and dermatology [15–18]. The second is full-field OCT, in which imaging is performed both in the reference and sample paths using lenses or microscope objectives [19–23]. Spatial and temporal coherence effects in interference microscopy and full-field optical coherence tomography were investigated in [24].

Recently, the authors presented work on the modulation of the degree of spatial coherence (DOSC). The optical grating strongly correlates with the coherence volume in the proposed DOSC laser scanning confocal microscope [25].

In most biological tissues, the scattering of light inside the tissue is the dominant factor that limits the imaging penetration depth. Since the scattering becomes weaker at longer wavelengths [26], the use of a light source with longer wavelengths should significantly improve the penetration depth. The spectral shaping for non-Gaussian source spectra in optical coherence tomography was discussed in [27]. Stroboscopic ultrahigh-resolution full-field optical coherence tomography was investigated in [28], and the effects of spatial coherence in diffraction phase microscopy were investigated in [29, 30].

The goal of the present work was to investigate the spatial coherence in confocal imaging systems using B/W concentric annuli for the 2nd microscope objective. We consider the detector to gain energy to a certain extent compared with the point detector. The spatial coherence factor (SCF) is deduced from the given formula in the theoretical analysis, and we obtain an empirical formula relating to the spatial coherence factor and the two-point objects. In addition, we obtained spatial coherence via the MATLAB method. Finally, we discuss the results and provide a conclusion.

The annular aperture has better resolution than the circular and quadratic apertures. In this chapter, a compromise of resolution and contrast is attained using the B/W concentric annulus since more intensity is attained compared with the annular aperture.

10.2 Theoretical Analysis

In this section, we calculate the mutual coherence factor using B/W concentric annular apertures for the confocal microscope objectives and consider the finite size of the detector aperture.

In the confocal arrangement shown in Fig. 10.1, the object is placed in the common short focus of the two objectives arranged in tandem. The apertures P_1 and P_2 have concentric B/W annuli, while the detector has a diameter of 2a. Coherent illumination is provided by a laser beam, while partially coherent detection is considered using a finite-size detector instead of a point detector.

It is known that the image of a point object is computed from the modulus square of the (RPSF). The complex amplitude at the detection plane is expressed in integral form as follows:

$$A(x, y) = \int\int_{-\infty}^{\infty} g(x_s, y_s) \cdot h_r(x - x_s, y - y_s) dx_s dy_s \tag{10.1}$$

Equation (10.1) represents a convolution product of the object $g(x, y)$, and the RPSF.

$$h_r(x, y) = h_1(x, y) \cdot h_2(x, y) \tag{10.2}$$

$h_1(x, y)$ represents the PSF corresponding to the 1st objective, while $h_2(x, y)$ represents the PSF corresponding to the 2nd objective. We write the PSF corresponding to the 1st objective of the aperture $P_1(u, v)$ as follows:

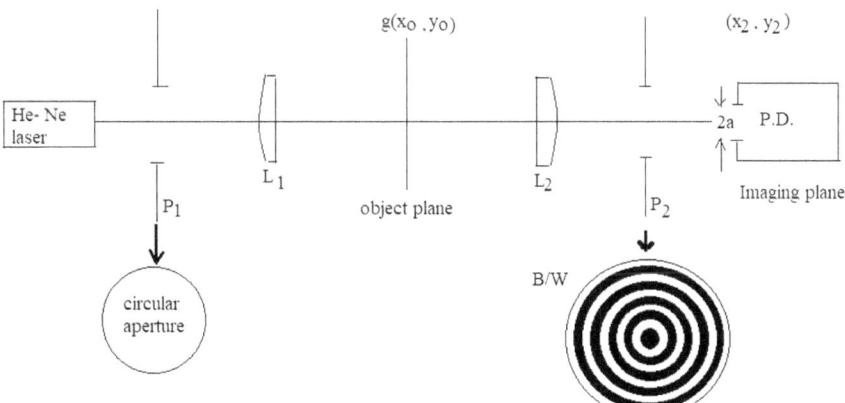

Fig. 10.1 Confocal optical system provided with a circular aperture for L_1 and B/W concentric annuli for L_2. The B/W aperture has 5 transparent zones, and the diameter of the detector $=$ is 2a

$$h_1(x, y) = \int\limits_{-\infty}^{\infty} \int P_1(u, v) \cdot \exp\left[\left(\frac{i2\pi}{\lambda f}\right)(ux + vy)\right] du dv \qquad (10.3)$$

A similar expression is given for $h_2(x, y)$.

The convolution of $g(x, y)$ and $h_r(x, y)$ is realized by object scanning in the (x, y) plane.

The detected intensity is the modulus square of Eq. (10.1) and is rewritten symbolically in radial coordinates $r = \sqrt{x^2 + y^2}$ as follows:

$$I(r) = |g(r) \otimes h_r(r)|^2 \qquad (10.4)$$

$\otimes$ is a symbol for convolution.

For a point object represented by a Dirac delta function (r) located at the center along the optical axis, Eq. (10.4) becomes:

$$I(w) = |h_1(w) \cdot h_2(w)|^2 \qquad (10.5)$$

The reduced coordinate $w = \frac{2\pi}{\lambda f}\rho_0 r$ replaces the radial coordinate r. ρ_0 represents the aperture radius, while f is the focal length corresponding to the symmetric objective lenses L_1 and L_2,

Now, consider an object composed of two luminescent points located symmetrically at points $x = \pm x_0$, $y = \pm y_0$, which is represented mathematically in radial coordinates as follows:

$$g(r) = \delta(r - r_0) + \delta(r + r_0) \qquad (10.6)$$

The radial distance between two points is $d = 2r_0$. By substituting Eq. (10.6) into Eq. (10.4) and making use of the linearity property of the Fourier transform and convolution operation, we finally obtain:

$$I(x, y) = |h_r(x - x_0, y - y_0) + h_r(x + x_0, y + y_0)|^2 \qquad (10.7)$$

$h_r(x, y)$ is computed from Eq. (10.2).

The intensity distribution for the two-point objects or the intensity impulse response calculated from Eq. (10.7) assumes that both illumination and detection are coherent. Since coherent detection requires a point detector that corresponds to a very weak intensity in the detection plane, a reasonable size is necessary for detection. Consequently, a confocal scanning microscope, which has coherent illumination and partially coherent detection, is investigated in the following section.

For stationary objects, $g(x_0, y_0)$, the complex amplitude in the imaging plane, considering a detector of finite size 2a, is calculated as [4]:

$$A_{st}(x_2, x_2) = \int\int_{-\infty}^{\infty} \left[h_1(x_0, y_0)g(x_0, y_0) \right] h_2(x_0 + x_2, y_0 + y_2)dx_0 dy_0 \qquad (10.8)$$

For a scanned object composed of two points, the complex amplitude formed in the imaging plane is obtained as follows:

$$A_{scan}(x_s, x_s) = \left[h_1(x_s - x_0, y_s - y_0) \right] h_2(x_s + x_2 - x_0, y_s + y_2 - y_0)$$
$$+ \left[h_1(x_s + x_0, y_s + y_0) \right] h_2(x_s + x_2 + x_0, y_s + y_2 + y_0) \qquad (10.9)$$

where (x_s, y_s) are the Cartesian scanning coordinates corresponding to the radial coordinate $r_s = \sqrt{x_s^2 + y_s^2}$.

The detected intensity distribution for the scanned object is obtained by taking the modulus square of the complex amplitude $A_{scan}(x_2, x_2)$, and then integrating over the surface area of the detector, that is:

$$I(x_s, y_s; a) = \int\int_{-\infty}^{\infty} D(x_2, y_2) |A_{scan}(x_2, x_2)|^2 dx_2 dy_2 \qquad (10.10)$$

$D(x_2, y_2)$ is the sensitivity of the detector with a circular aperture of radius a. The detector aperture is represented in the radial coordinate $r_2 = \sqrt{x_2^2 + y_2^2}$ as follows:

$$D(r_2) = 1, \quad \text{when } \frac{r_2}{a} \leq 1 \qquad (10.11)$$

Substituting Eq. (10.11) with Eq. (10.10), we finally obtain:

$$I(r_s; a) = h_1^2(r_s - r_0) \int_0^a h_2^2(r_s + r_2 - r_0)r_2 dr_2$$
$$+ h_1^2(r_s + r_0) \int_0^a h_2^2(r_s + r_2 + r_0)r_2 dr_2$$
$$+ 2h_1(r_s - r_0)h_1(r_s + r_0)\gamma_{12} \qquad (10.12)$$

where γ_{12} is the mutual coherence function for the scanned object and is computed as follows:

$$\gamma_{12} = \int_0^a h_2(r_s + r_2 - r_0) \cdot h_2^*(r_s + r_2 + r_0)r_2 dr_2 \qquad (10.13)$$

For stationary objects, $r_s = 0$. The mutual coherence function is:

$$\gamma_{12} = \int_0^a h_2(r_2 - r_0) \cdot h_2^*(r_2 + r_0)r_2 dr_2 \qquad (10.14)$$

Now, the PSF h_1 and h_2 are computed as follows: The mutual coherence function is dependent upon PSF h_2 corresponding to the objective used for detection.

We assume that pupil P_1 corresponding to the 1st objective has a uniform circular distribution, while pupil P_2 corresponding to the 2nd objective has a B/W concentric annular distribution, as shown in Fig. 10.1. Hence, we compute the point spread function (PSF) corresponding to each objective lens.

The PSF for a circular aperture is represented by the known Airy disk as follows:

$$h_1(w) = F.T.\{P_1(\rho)\} = \frac{2J_1(w)}{w} \tag{10.15}$$

where $w = 2\pi \frac{\rho_0 r}{\lambda f}$ is the reduced coordinate, r is the radial coordinate located in the focal plane f corresponding to the 1st objective lens L_1, and ρ_0 represents the aperture radius.

The PSF corresponding to the 2nd aperture P_2, which has B/W concentric annuli, is obtained as follows:

$$h_2(w) = 2\pi\rho_{max}^2 \sum_{i=1}^{N} \left\{ \frac{2J_1(w_{2i-1})}{w_{2i-1}} - \epsilon_i^2 \frac{2J_1(w_{2i})}{w_{2i}} \right\} \tag{10.16}$$

where $\epsilon_i = \frac{\rho_{2i}}{\rho_{2i-1}}$; $i = 1, 2, \ldots, 5$.

A limited number of transparent zones $N = 5$ is assumed, as shown in Fig. 8.2, and $\rho_1 = \rho_{max}$ is for pupil P_1, etc.

Substituting Eq. (10.16), which represents the PSF corresponding to the 2nd objective, into Eq. (10.14), we finally obtain:

$$\gamma_{12} = \left(4\pi^2 \rho_{max}^4\right) \sum_{i=1}^{N} \int_0^a \left\{ \frac{J_1(w_{2i-1})}{w_{2i-1}} - \epsilon_i^2 \frac{J_1(w_{2i})}{w_{2i}} \right\}$$

$$\cdot \left\{ \frac{J_1(w'_{2i-1})}{w'_{2i-1}} - \epsilon_i^2 \frac{J_1(w'_{2i})}{w'_{2i}} \right\} r_2 \, dr_2$$

$$w_{2i-1} = \frac{2\pi\rho_{2i-1}(r - r_0)}{\lambda f} \quad , \quad w'_{2i-1} = \frac{2\pi\rho_{2i-1}(r + r_0)}{\lambda f} \quad , \tag{10.17}$$

$w_{2i} = \frac{2\pi\rho_{2i}(r - r_0)}{f}$, and $w'_{2i} = \frac{2\pi\rho_{2i}(r + r_0)}{f}$ represents the reduced coordinates corresponding to the defined zones in Eq. (10.17).

For an annular aperture, only two concentric circles are kept where the internal circle is dark. Hence, the spatial coherence function γ_{12} in Eq. (9.17) is reduced to:

$$\gamma_{12} = \left(4\pi^2\rho_{max}^4\right) \int_0^a \left[\frac{J_1(w_1)}{w_1} - \epsilon_1^2 \frac{J_1(w_2)}{w_2} \right] \left[\frac{J_1(w'_1)}{w'_1} - \epsilon_1^2 \frac{J_1(w'_2)}{w'_2} \right] r_2 \, dr_2;$$

$$\epsilon_1 = \frac{\rho_2}{\rho_1} \tag{10.18}$$

We consider the following reduced coordinates in Eq. (10.18):

$$w_1 = \frac{2\pi\rho_1(r - r_0)}{f}, \; w_2 = \frac{2\pi\rho_2(r - r_0)}{f},$$

$$w'_1 = \frac{2\pi\rho_1(r + r_0)}{f}, \; \text{and } w'_2 = \frac{2\pi\rho_2(r + r_0)}{f}$$

For the circular aperture, we obtain:

$$\gamma_{12} = \left(4\pi^2\rho_{max}^4\right) \int_0^a \frac{J_1(w_1)}{w_1} \frac{J_1(w'_1)}{w'_1} r_2 dr_2 \tag{10.19}$$

The above integrals in Eqs. (10.17), (10.18), and (10.19) are solved based on numerical integration using MATLAB.

The relation of γ_{12} versus the aperture radius of the detector a is represented by an empirical formula as follows:

$$\gamma_{12}(a; r_0) = \beta\left(1 - e^{-\alpha a}\right) \tag{10.20}$$

where β is a constant to be determined from the computed results and $a = 1$ has the units μm^{-1} This empirical formula is valid for apertures that are either circular or annular.

10.3 Results and Discussion

The coherence function γ_{12} computed from Eqs. (10.17), (10.18), and (10.19) is dependent upon the detector size and the distance r_0 between the two-point objects for a certain NA corresponding to the objective lens placed in front of the detector.

The SCF vs. the distance between the two-point objects r_0 for three different apertures with detector radius $a = 1$ mm is shown in Fig. 10.2. MATLAB code is used in the computation, where the continuous curve represents the B/W aperture and is compared with the circular and annular apertures. The SCF curve decreases without fringing in the case of B/W concentric annuli.

Another plot of the SCF or γ_{12} versus r_0 for B/W concentric annuli is shown in Fig. 10.3 for detector apertures of sizes $a = 0.6, 0.8$, and 1.0 mm. The SCF decreases as the distance between the two-point objects increases.

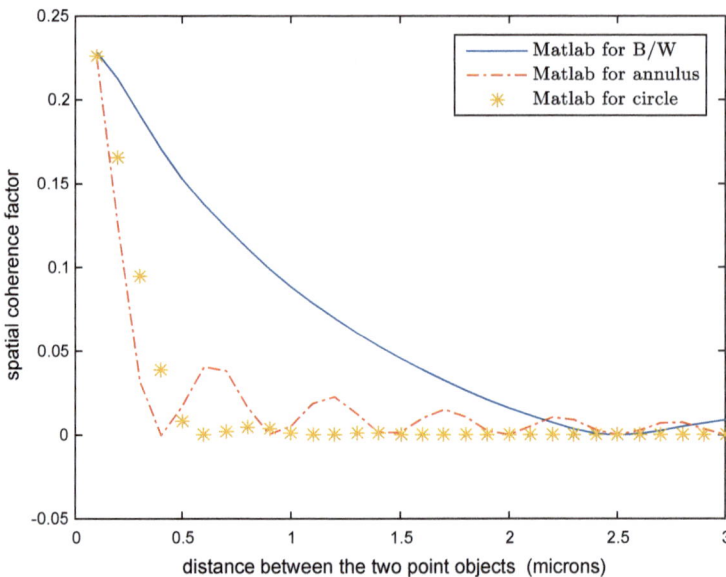

Fig. 10.2 The SCF versus the distance between the two-point objects r_0 for three different apertures with detector radius $a = 1$ mm. MATLAB is used in the computation where the continuous curve represents the D/W aperture and is compared with the circular and annular apertures

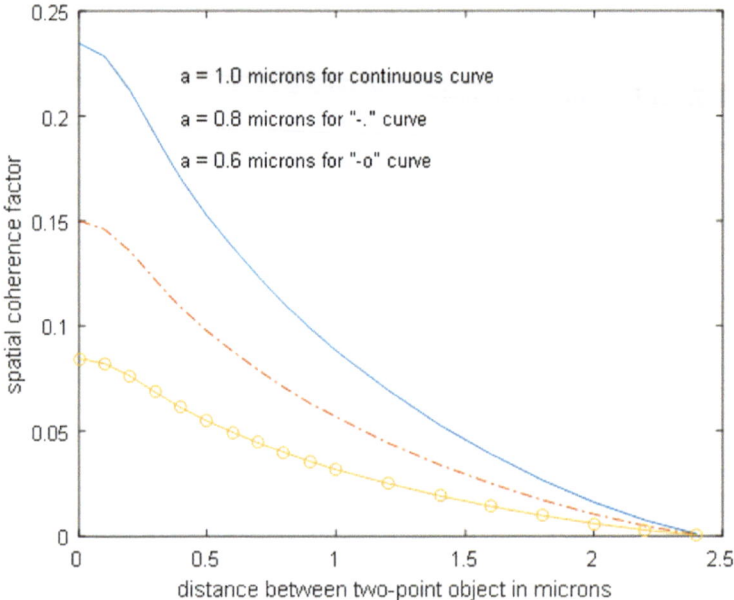

Fig. 10.3 The spatial coherence function vs. the distance r_0 with different radii of $a = 0.6, 0.8,$ and 1.0 mm for the detector. A B/W aperture of NA $_{external} = 0.6$ and 5 transparent equal zones is considered

A plot of γ_{12} versus the detector size is shown in Fig. 8.4 for three different values of $r_0 = 0.1, 0.2$, and 0.3 mm. γ_{12} increases exponentially with the detector radius, which is represented by the empirical formula (10.20).

The plots shown in Figs. 10.3 and 10.4 for the B/W transparent concentric annuli are compared with the results shown in Figs. 10.5 and 10.6 for the uniform circular aperture showing similarity in γ_{12}. The results corresponding to the annular aperture, represented in Fig. 10.7, showed fringing at $r_0 > 0.4$ mm. This is attributed to the appearance of sharp legs in the PSF compared with that corresponding to the circular aperture. The plot of γ_{12} versus the detector radius follows the empirical exponential formula (10.20), where the plot is shown in Fig. 10.8.

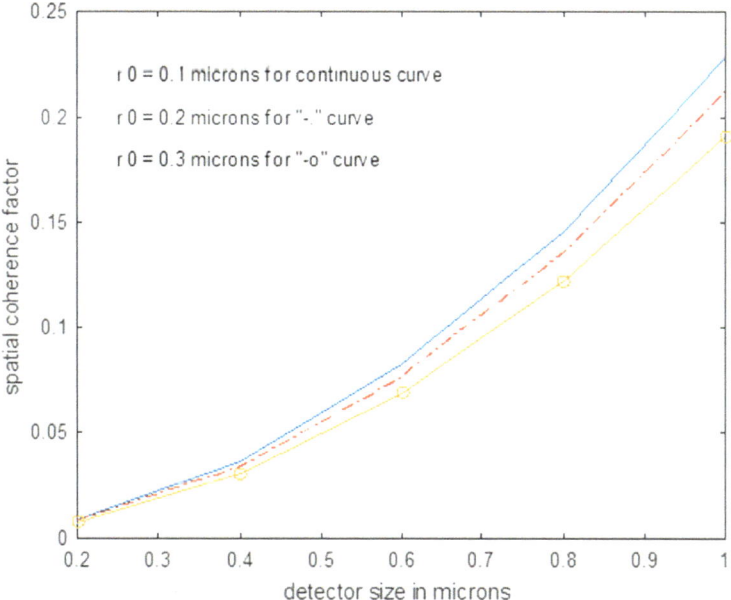

Fig. 10.4 The spatial coherence function versus the detector size a using two-point objects of $r_0 = 0.1, 0.2,$ and 0.3 mm. B/W concentric annuli, of 5 transparent zones, are considered

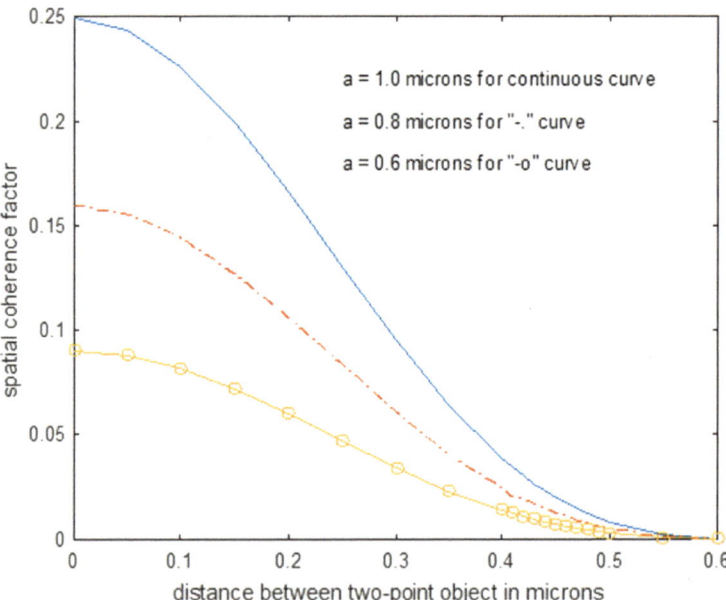

Fig. 10.5 The spatial coherence function vs. the distance r_0 using different radii for the detector of $a = 0.6,\ 0.8,$ and 1.0 mm. The circular aperture of NA $= 0.6$ is considered

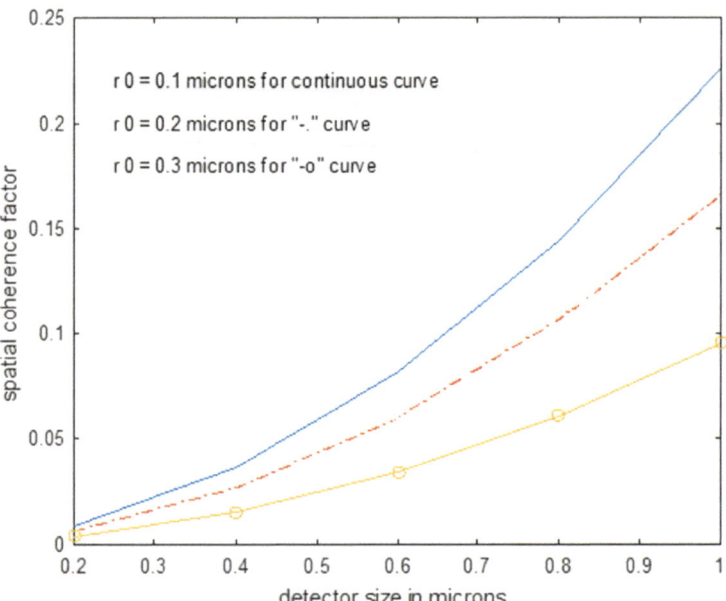

Fig. 10.6 The spatial coherence function vs. the detector size a using two-point objects of $r_0 = 0.1, 0.2,$ and 0.3 mm. The circular aperture of NA $= 0.6$ is considered

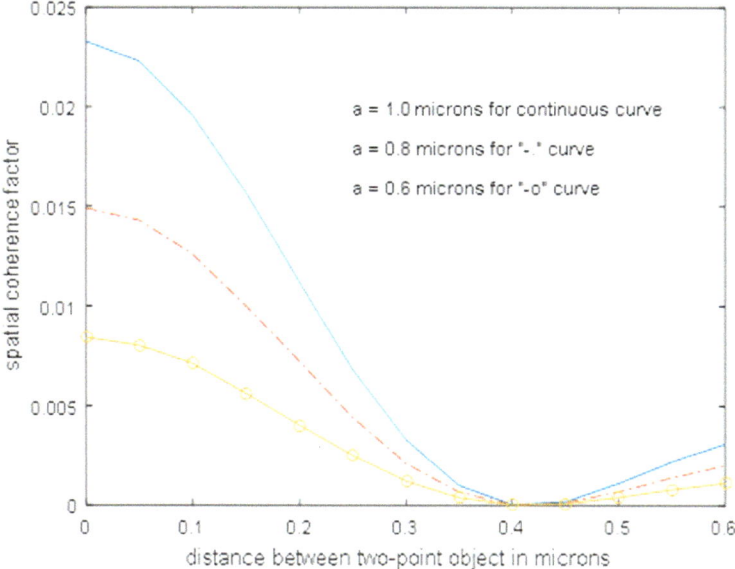

Fig. 10.7 The spatial coherence function vs. the distance r_0 using different radii for the detector of $a = 0.6$, 0.8, and 1.0 mm. The annular aperture of NA $_{external} = 0.6$ and NA $_{internal} = 0.5$ is considered

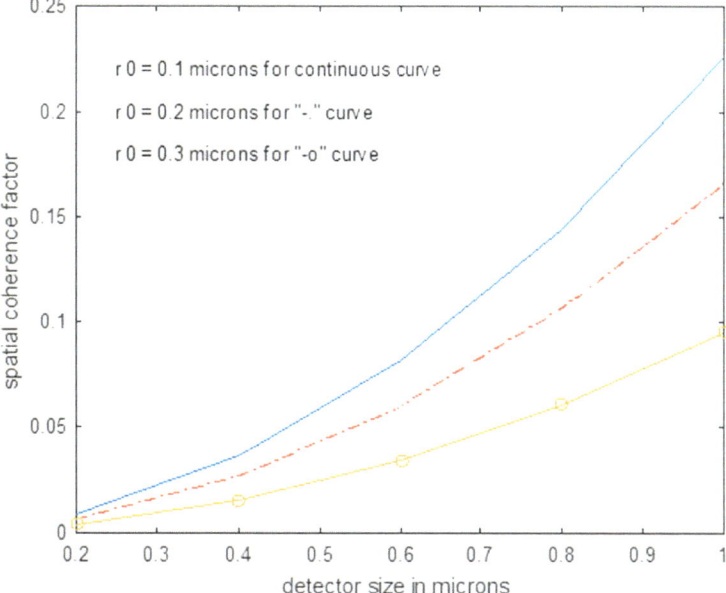

Fig. 10.8 The spatial coherence function vs. the detector size a using two-point objects of $r_0 = 0.1$, 0.2, and 0.3 mm. The annular aperture is considered

10.4   Conclusion

First, we computed the SCF versus the two-point objects using B/W concentric annuli and compared it with that corresponding to the circular and annular apertures.

Second, in the case of an annular aperture, the coherence factor is cut off at $r_0 = 0.4$ mm. The fringing in the coherence factor is attributed to the appearance of legs in the diffraction pattern or the PSF at $r_0 > 0.4$ mm.

The obtained results of SCF versus the distance between the two points are represented by an empirical formula showing exponential variation fitted with theoretical calculations.

Finally, a compromise of resolution and contrast is attained using the B/W concentric annulus since more intensity is attained compared with the annular aperture, which is useful in confocal microscopic imaging.

References

1. M. Francon, S. Slansky, *Coherence en optique*, Fascicule II (1963)
2. G.W. Goodman, *Introduction to Fourier Optics* (McGraw Hill, NY, 1968)
3. C.J.R. Sheppard, Image formation in scanning microscopes with partially coherent source and detector. Opt. Acta **25**, 315–325 (1978)
4. C.J.R. Sheppard, T. Wilson, Multiple traversing of the object in the scanning microscope. Opt. Acta **27**, 611–624 (1980)
5. C.J.R. Sheppard, Image formation in confocal scanning microscopes. Optik **55**, 331–342 (1980)
6. A.M. Hamed, Optimization of spatial coherence in confocal optical systems. Opt. Laser Tech. **22**, 137–139 (1990). https://doi.org/10.1016/0030-3992(90)90024-X
7. A.M. Hamed, A study on spatial coherence using quadratic radially distributed apertures (application to confocal imaging). Opt. Laser Tech. **29**, 93–95 (1997). https://doi.org/10.1016/S0030-3992(96)00003-5
8. A.M. Hamed, J.J. Clair, Image and super-resolution in optical coherent microscopes. Optik **64**, 277–284 (1983)
9. D. Farkas, J.G. Fujimoto, *Biomedical Optical Imaging*. Chapter 1: Confocal microscopy by T. Wilson (Oxford University Press, 2009)
10. J.B. Pawley, *Handbook of the Biological Confocal Microscope* (Plenum Press, New York, 2000)
11. B. Sick, B. Hecht, L. Novotny, Orientational imaging of single molecules by annular illumination. Phys. Rev. Lett. **85**, 4482–4485 (2000)
12. M.A.A. Neil, M.J. Booth, T. Wilson, New modal wave-front sensor: a theoretical analysis. J. Opt. Soc. Am. **17**, 1098–1107 (2000)
13. W. Gao, Effects of temporal and spatial coherence on resolution in full-field optical coherence tomography. J. Mod. Opt. **62**, 1764–1774 (2015)
14. S.W. Hell, E.M.K. Stelzer, S. Lindek, C. Cremer, Confocal microscopy with an increased detection aperture: type-B 4Pi confocal microscopy. Opt. Lett. **19**, 222 (1994)
15. D. Huang et al., Optical coherence tomography. Science **254**, 1178–1181 (1991)
16. B.E. Bouma, G.J. Tearney (eds.), *Handbook of Optical Coherence Tomography* (Marcel Dekker, New York, 2002)
17. J.R. Wilkins, Characterization of epiretinal membranes using optical coherence tomography. Ophthalmology **103**, 2142–2151 (1996)

18. J. Welzel, E. Lankenau et al., Optical coherence tomography of the human skin. J. Am. Acad. Dermatol. **37**, 958–963 (1997)
19. E. Beaurepaire, A.C. Boccara et al., Full-field optical coherence microscopy. Opt. Lett. **23**, 244–246 (1998)
20. A. Dubois, Laurent Vabre et al., High-resolution full-field optical coherence tomography with a Linnik microscope. Appl. Opt. **41**(4)**,** 805–812 (2002)
21. A. Dubois et al., Ultrahigh-resolution full-field optical coherence microscopy. Appl. Opt. **43**, 2874–2883 (2004)
22. W.Y. Oh et al., Ultrahigh-resolution full-field optical coherence microscopy using InGaAs camera. Opt. Express **14**, 726–735 (2006)
23. B. Laude et al., Full-field optical coherence tomography with thermal light. Appl. Opt. **41**, 6637–6645 (2002)
24. Abdulhalim, Spatial and temporal coherence effects in interference microscopy and full-field optical coherence tomography. Ann. Phys. (Berlin) **524**, 787–804 (2012). https://doi.org/10.1002/andp.201200106.
25. Chen, C.-C. et al., Degree-of-spatial-coherence laser scanning confocal fluorescence microscope. Opt. Commun. **518** (2022). https://doi.org/10.1016/j.optcom.2022.128315
26. E. Bordenave, E. Abraham et al., Wide-field optical coherence tomography: imaging of biological tissues. Appl. Opt. **41**, 2059–2064 (2002)
27. R. Tripathi, N. Nassif et al., Spectral shaping for non-Gaussian source spectra in optical coherence tomography. Opt. Lett. **27**, 406–408 (2002)
28. G. Moneron, A.C. Boccara, A. Dubois, Stroboscopic ultrahigh-resolution full-field optical coherence tomography. Opt. Lett. **30**, 1351–1353 (2005)
29. T.H. Nguyen, G. Popescu, Spatial Light Interference Microscopy (SLIM) using twisted-nematic liquid-crystal modulation. Biomed. Opt. Express **4**(9), 1571–1583 (2013)
30. C. Edwards et al., Effects of spatial coherence in diffraction phase microscopy. Opt. Express **22**(5), 5133–5146 (2014)

Chapter 11
Application of Confocal Laser Scanning Microscopy to Cardiac Images

We suggested the use of a cardioid aperture in the processing of cardiac images via confocal laser scanning microscopy (CSLM). We calculated the point spread function (PSF) corresponding to the cardioid aperture and the cardiac images by using the FFT algorithm. Then, we studied the effect of image rotation on the PSF. In addition, we computed the autocorrelation corresponding to the cardiac images and cardiac aperture and compared them with the autocorrelation of the uniform circular aperture. Additionally, we computed the cross-correlation of the aligned and rotated cardiac images. Finally, we obtained cardiac images using a confocal scanning microscope with circular and cardioid apertures. The MATLAB code was used for all images and plots.

11.1   Introduction

Digital image processing is the art of treating digital images [1] to obtain information from the image. The first research comes from the science of astronomy regarding stars and galaxies. These images provide results for medical imaging, which is a major concern of in vivo diagnosis instead of in vitro diagnosis. Imaging diseased organs can lead to a diagnosis without the need for a biopsy [2]. Biopsy hurts on the other side, and imaging is quite easy and simple. The digital image is presented as an image where the intensity in the image is quantized in 256×512 Gy levels. On the other hand, color images can be discussed.

The digital image is treated by algebraic and mathematical tools to obtain different information about the image. For example, digital differentiation can result in edge detection and the skeleton of objects in the image. Spatial and frequency filters can be applied to an image to obtain the edged image or to obtain the averaged image via a low-frequency pass. Fourier transforms (FTs) are computed using MATLAB, or they are imaged in space by a convex length. We can pass or reject different regions of

© The Author(s), under exclusive license to Springer Nature Switzerland AG 2025
A. M. Hamed, *Studies on the Confocal Laser Microscope*,
SpringerBriefs in Applied Sciences and Technology,
https://doi.org/10.1007/978-3-031-87275-4_11

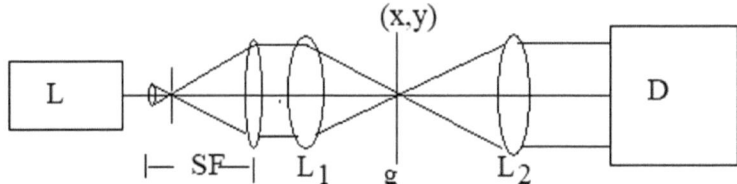

Fig. 11.1 Optical set-up for the confocal scanning laser microscope. L_1 and L_2 objective lenses correspond to illumination and detection, respectively. L: laser beam, SF: spatial filter, g: mechanically scanned object in the plane (x, y)., D: coherent detector

frequency of interest. Moreover, the morphological operation was found to provide different image manipulations to provide the necessary image processing techniques [3]. Different algorithms are applied to the image [4, 5].

Moreover, random processes faced in images and signal processing are random. Thus, the image is random. Image processing requires random image processing [6] or the use of a fuzzy technique [7].

A confocal scanning laser microscope consists of two objectives arranged in tandem, where the scanned object is placed in the common short focus corresponding to the two objectives, as shown in Fig. 10.1. The image is shown in the detection plane constructed where the electronic scanning is synchronized with the mechanical scanning of the object. This confocal microscope considers coherent illumination emitted from the laser beam and coherent detection. Consequently, the intensity corresponding to the image is the modulus square of the convolution product of the object and the resultant PSF. The resultant PSF is the simple product of the PSF corresponding to the two apertures. In this study, one aperture had a cardiac shape, while the other had a uniform circular shape (Fig. 11.1).

Recently, image processing of medical images based on the speckle imaging technique and its contrast was presented in [8–12], while image processing using a confocal scanning laser microscope provided with modulated apertures as outlined in [13–15]. Another recent work using confocal microscopy was presented in [16–18]. Compressive light-field microscopy for 3-D neural activity recording is described in [16]. The remodeling of cardiac tissue was analyzed, and a comprehensive approach based on confocal microscopy leading to 3D reconstruction was presented in [17]. Catheterized Fiber-Optics Confocal Microscopy of the Beating Heart in Situ was investigated in [18].

This work aimed to process cardiac images using a cardiac aperture in a confocal scanning laser microscope. The chapter is organized as follows. The second section presents an analysis of the computation of the point spread function (PSF) using a cardioid curve. Section 11.3 presents the results and discussion. Section 11.4 presents conclusions.

11.2 Analysis

The cardiac aperture in Cartesian coordinates is represented as follows [1, 2]:

$$p(x, y) = 1; \quad \frac{(x^2+y^2-2ax)^2}{4a(x^2+y^2)} \leq 1$$
$$= 0; \quad \text{otherwise} \tag{11.1}$$

In polar coordinates, it is represented as follows:

$$p(\rho, \theta) = 1; \rho(\theta) = 2a[1 - \cos(\theta)] \tag{11.2}$$

a is a constant, while the radius ρ changes with the angle θ.
 The constant a is deduced from Eq. (11.1) as follows:

$$a = \frac{x^2 + y^2}{2[x + \sqrt{x^2 + y^2}]} \tag{11.3}$$

Equation (11.3) only holds for the variables (x) and (y), which lie on the contour of the cardioid, and parametric equations of the planar cardioid are written as follows:

$$x = 2a\cos(\theta)\big[1 - \cos(\theta)\big], \quad \text{and } y = 2a\sin(\theta)\big[1 - \cos(\theta)\big] \tag{11.4}$$

The PSF for the cardiac aperture is computed by applying the Fourier transform in polar coordinates as follows:

$$h(r) = \int_0^{2\pi} \int_0^{\rho} P(\rho, \theta) \exp\left\{-\frac{j2\pi}{\lambda f}(\rho r \cos\theta)\right\} \rho \, d\rho \, d\theta \tag{11.5}$$

where ρ is a variable dependent on the angle θ, represented by Eq. (11.2), and r is the radial coordinate in the Fourier plane, where $r = \sqrt{u^2 + v^2}$.
 In the next section, we computed the PSF corresponding to the cardiac aperture Eq. (11.2) using the fast Fourier transform (FFT).

11.3 Results and Discussion

We computed the cardiac aperture from Eqs. (11.1) and (11.2), using MATLAB and plotted it as shown in Fig. 11.2. The radius ρ is not a constant, as in a circular aperture; it varies, according to Eq. (11.2), in the range $[0, 2\pi]$.

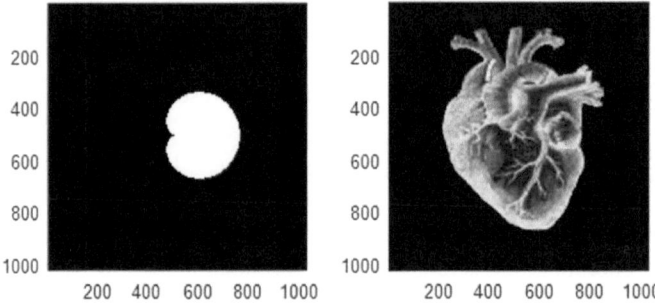

Fig. 11.2 In the L.H.S. Cardioid aperture is shown while in the R.H.S., the heart image is shown. The two images have dimensions of 1024 × 1024 pixels. The maximum radius for the aperture = 256 pixels while the max. radius for the heart image = 512 pixels

The cardioid aperture shown in the L.H.S. is assumed to be a model for comparison with the heart image shown in the R.H.S., as shown in Fig. 11.2. The two images have dimensions of 1024 × 1024 pixels. The average radius for the aperture is 128 pixels, while for the heart image, the average radius is 256 pixels. The cardioid aperture image was constructed using the formula for the cardioid curve [1].

The heart image rotation from $\theta = [0° \ 180°]$, where $\theta = 15°$ is shown in Fig. 11.3, while other image rotations are evident.

The normalized PSF corresponding to the aperture of the cardioid shape, using the FFT at a maximum radius = 32 pixels, is computed and plotted, as shown in Fig. 11.4a. The curve shows that the FWHM = 15 pixels. It is computed for the cardiac image and plotted as shown in Fig. 11.4b. A comparative plot of the normalized PSF for a circular aperture with a radius = 32 pixels is shown in Fig. 11.4c. The FWHM is

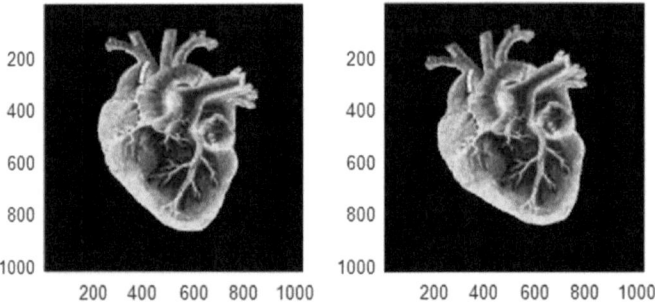

Fig. 11.3 In the L.H.S. heart image is aligned along the Cartesian coordinates shown while in the R.H.S., the heart image is inclined with an angle $\theta = 15°$

shown as 20 pixels. The heart image PSF is narrower than the PSF corresponding to the cardiac and circular apertures. In addition, the PSF has an asymmetric distribution, which is attributed to the heart geometry compared with the symmetric shapes for the cardiac and circular apertures. The stronger legs in Fig. 11.4a, c compared with the weaker legs in Fig. 11.4b correspond to the PSF plot of the heart image. Hence, the resolution attained for heart images is better than the resolution obtained for cardiac and circular apertures. By comparing the PSF corresponding to the circular and cardioid apertures, we obtained a narrower FWHM for the cardioid aperture than for the circular aperture, as shown in Fig. 11.4a, c.

The PSF corresponds to the rotated images computed and plotted as shown in the Fig. 11.5a, b, …. The figure shows the asymmetric distribution for the PSF corresponding to image rotation. In addition, the PSF is sensitive to image rotation since the shape modified in the PSF ranges from 0 to 0.3 around the peak. It is interesting to show the same profile for the PSF for the unrotated image shown in Fig. 11.4b and its rotated image with $\theta = 180°$ shown in Fig. 11.5f. The normalized PSF for the inclined cardiac image at an angle $\theta = 5°$ shown in Fig. 11.5a.

The autocorrelation function of the aperture of the cardioid shape was computed and plotted in Fig. 11.6. A line plot corresponding to the autocorrelation of the cardioid model at $y = 512$ pixels is shown in Fig. 11.7a. The calculated FWHM $= 81$ pixels. The line plot at $x = 512$ pixels is shown in Fig. 11.7b. FWHM $= 68$ pixels. Hence, unequal FWHM values are attributed to the elongation of the cardioid aperture along one of the axes to the other.

The autocorrelation intensity of the heart image is shown in Fig. 11.8. It shows a luminous spot at the center surrounded by a blurred image, as expected from the geometry of the heart image with its arteries. The line plot corresponds to the auto-correlation of the heart at $x = 512$ pixels shown in Fig. 11.9a and the corresponding FWHM $= 307$ pixels, while the autocorrelation at $y = 512$ pixels shown in Fig. 11.9b and the corresponding FWHM $= 403$ pixels. The figure shows an irregular distribution originating from the shape of the heat image in both plots. The cross-correlations of the aligned and rotated cardiac images were computed and plotted, for rotation angles $\theta = 30, 60,$ and $90°$, as shown in Fig. 11.10a. The cross-correlation plots for rotation angles $\theta = 120°, 150°,$ and $180°$ are shown in Fig. 11.10b. The results show different shapes for cross-correlation depending on the angle of rotation. It shows an asymmetric shape compared with the symmetric autocorrelation of the aligned cardiac image.

The cardiac and circular apertures used in the confocal microscopy images are shown in Fig. 11.11a. The maximum radius for the cardiac aperture equals the circular radius $= 64$ pixels. The original cardiac image and its reconstructed image using cardiac and circular apertures are shown in Fig. 11.11a and are plotted in Fig. 11.11b, while two symmetric circular apertures are shown in Fig. 11.11c. All images have the same matrix dimensions of 512×512 pixels.

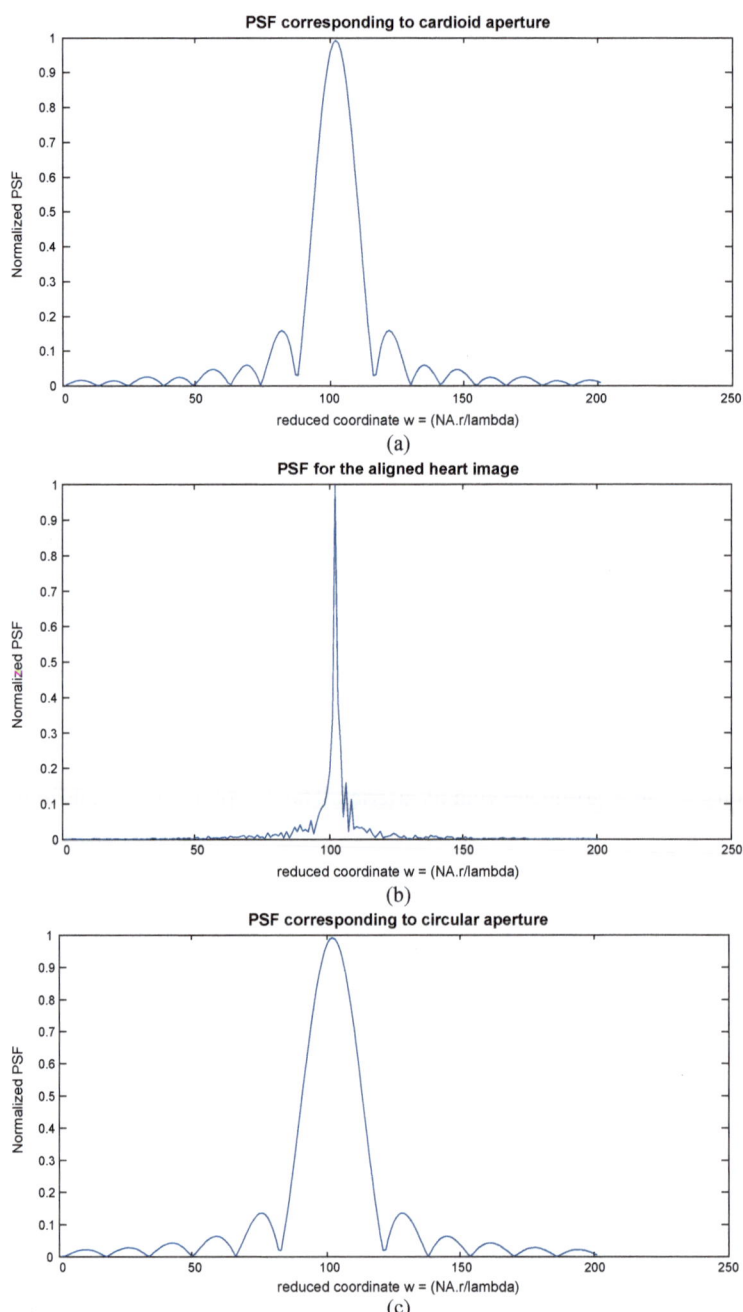

Fig. 11.4 a Normalized PSF corresponding to the aperture of the cardioid shape with a maximum radius = 32 pixels. The FWHM is equal to 15 pixels. **b** Normalized PSF corresponding to the cardiac image. The FWHM has an asymmetric distribution, which is attributed to heart geometry. **c** Normalized PSF for a circular aperture of radius = 32 pixels. The FWHM is shown as 20 pixels

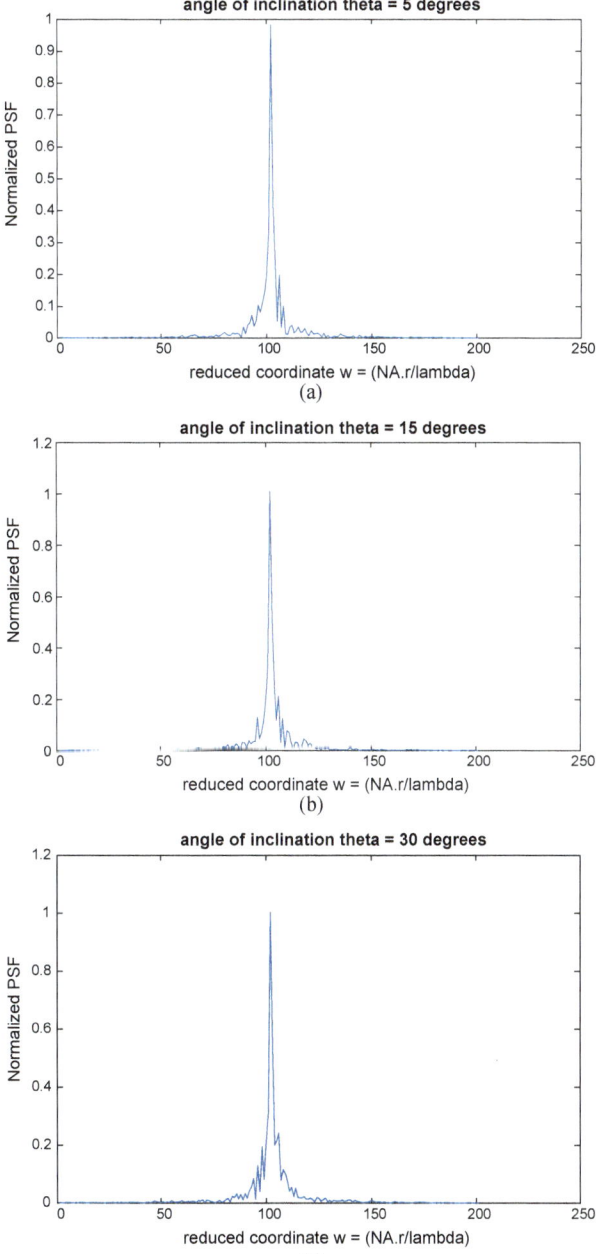

◀**Fig. 11.5   a** The normalized PSF for the inclined cardiac image at an angle $\theta = 5°$. **b** The normalized PSF for the inclined cardiac image at an angle $\theta = 15°$. **c** The normalized PSF for the inclined cardiac image at an angle $\theta = 30°$. **d** The normalized PSF for the inclined cardiac image at an angle $\theta = 45°$. **e** The normalized PSF for the inclined cardiac image at an angle $\theta = 60°$. **f** The normalized PSF for the inclined cardiac image at an angle $\theta = 90°$. **g** The normalized PSF for the inclined cardiac image at an angle $\theta = 120°$. **h** The normalized PSF for the inclined cardiac image at an angle $\theta = 150°$. **i** The normalized PSF for the inclined cardiac image at an angle $\theta = 180°$

11.4   Conclusion

First, the heart image is compared with the aperture of the geometric cardioid. It shows a symmetric PSF shape in the case of the aperture compared with the asymmetric curve in the case of the heart image. The FWHM computed from the PSF for the heart image is narrower than the corresponding FWHM for the cardioid model and circular apertures. In addition, heart image rotation showed different PSF profiles in the range ≤ 0.3 around the peak of different irregularities.

Second, the aperture of the cardioid shape has a narrower FWHM = 15 pixels compared with the FWHM of 20 pixels for the circular aperture considering the same radius of 32 pixels. Hence, we showed that the cardioid aperture may provide better results than the conventional circular aperture when used in a confocal set-up.

Third, the autocorrelation corresponding to the heart image along the Cartesian coordinates showed different values for the FWHM. This is attributed to the different dimensions due to image elongation. In addition, it has an irregular symmetric distribution depending on the image shape. The comparative cardioid aperture showed a triangular shape. The cross-correlation of the aligned and rotated cardiac images showed an irregular asymmetric distribution.

Finally, the reconstructed images obtained using a confocal microscope showed moderately resolved images dependent on the PSF for the cardioid aperture compared with the PSF corresponding to the ordinary circular apertures.

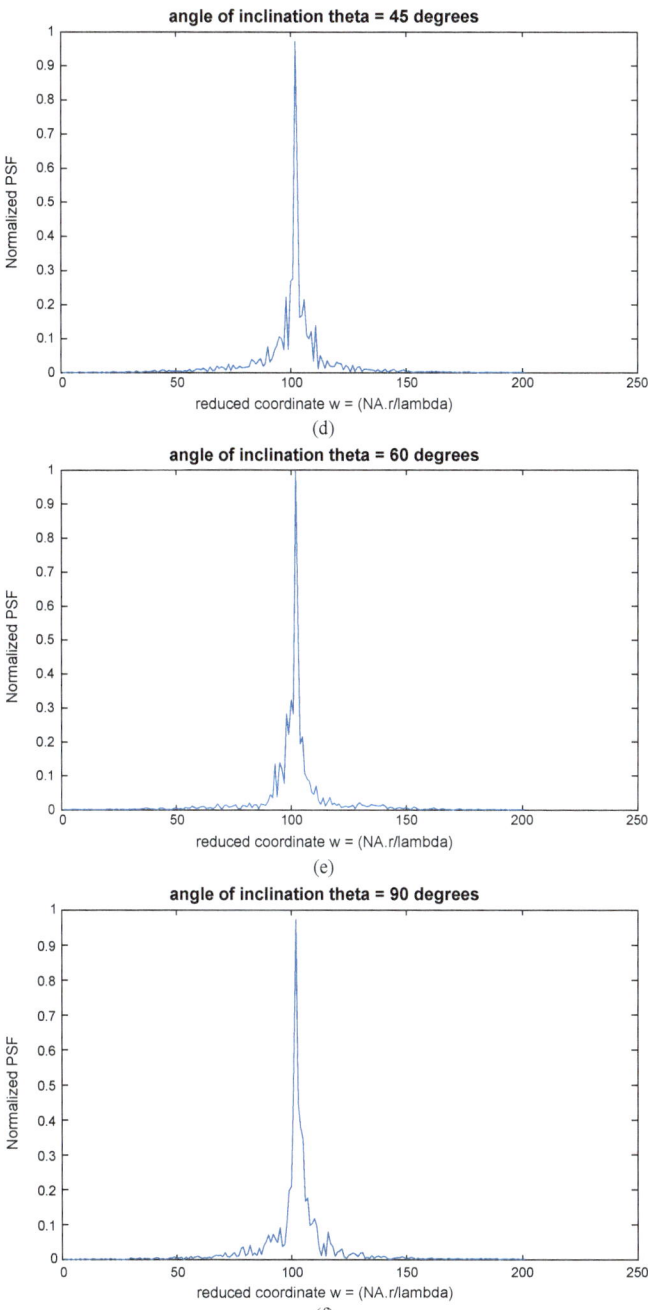

Fig. 11.5 (continued)

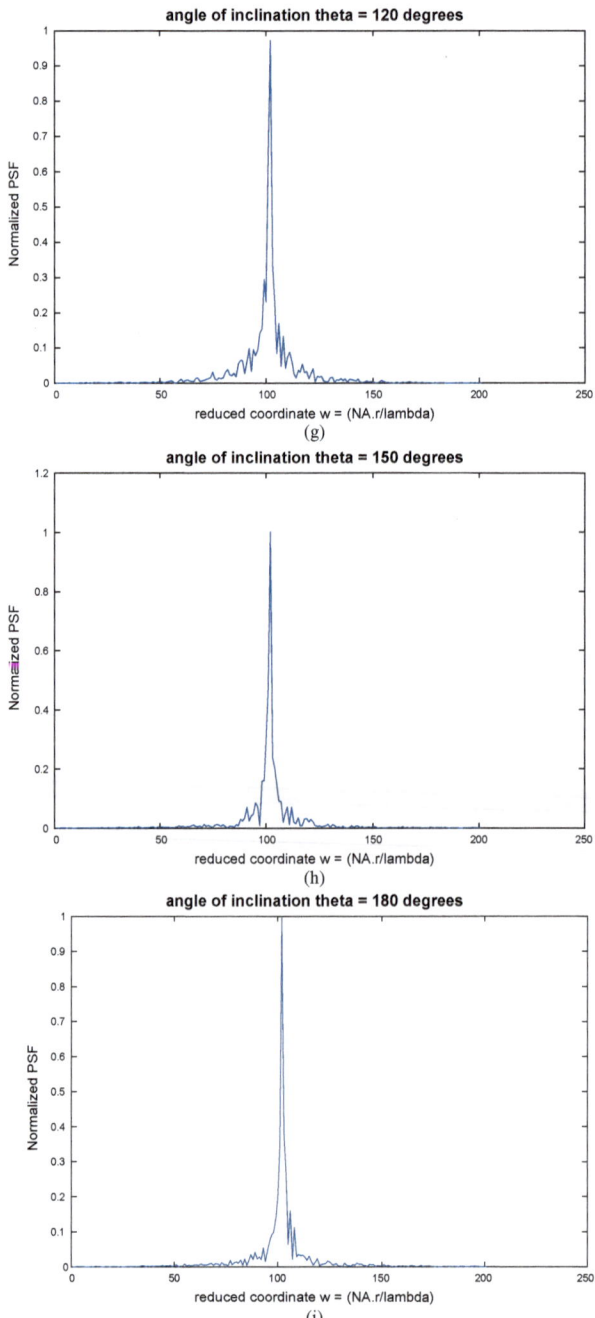

Fig. 11.5 (continued)

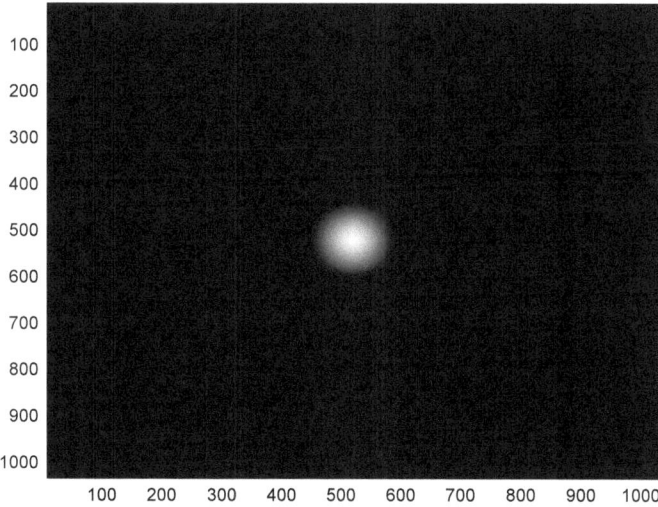

Fig. 11.6 Autocorrelation of the aperture of the cardioid shape shown in Fig. 11.2

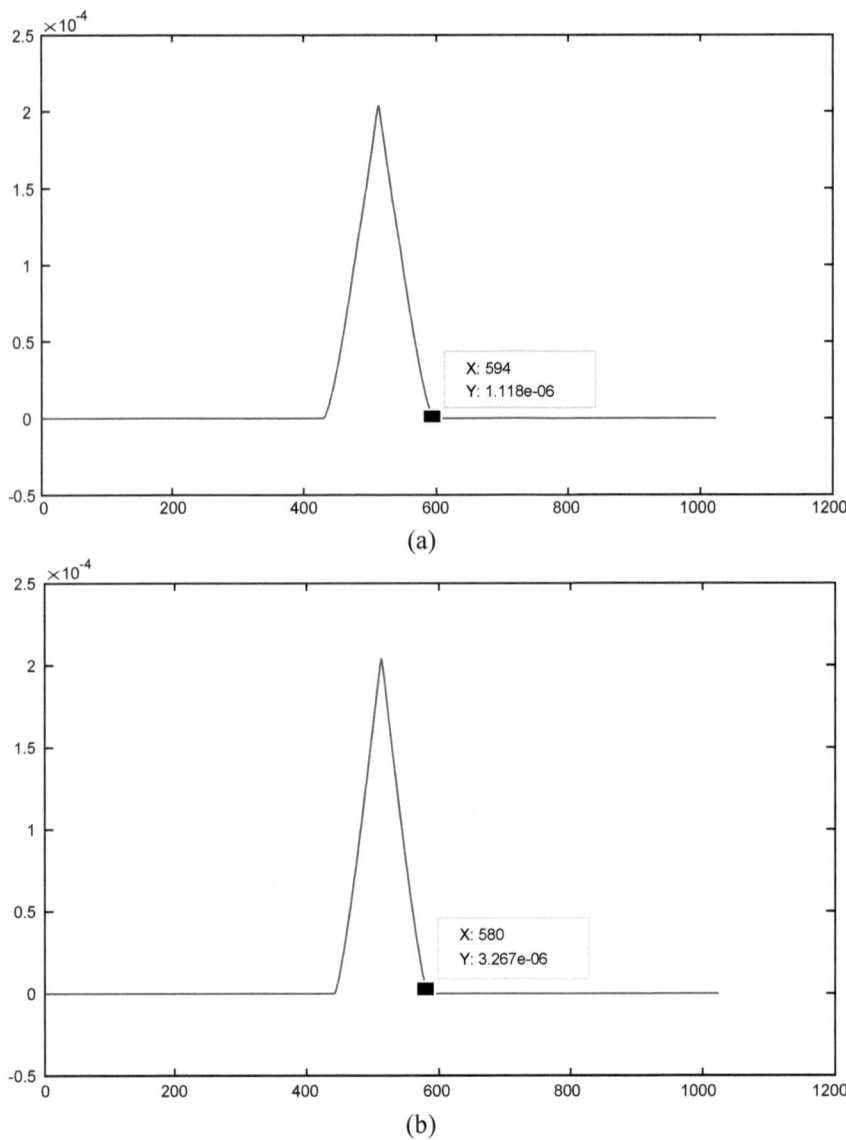

Fig. 11.7 **a** Line plot corresponding to the autocorrelation of the cardioid model at $y = 512$ pixels. The FWHM $= 81$ pixels. **b** Line plot corresponding to the autocorrelation of the cardioid model at $x = 512$ pixels. The FWHM $= 68$ pixels

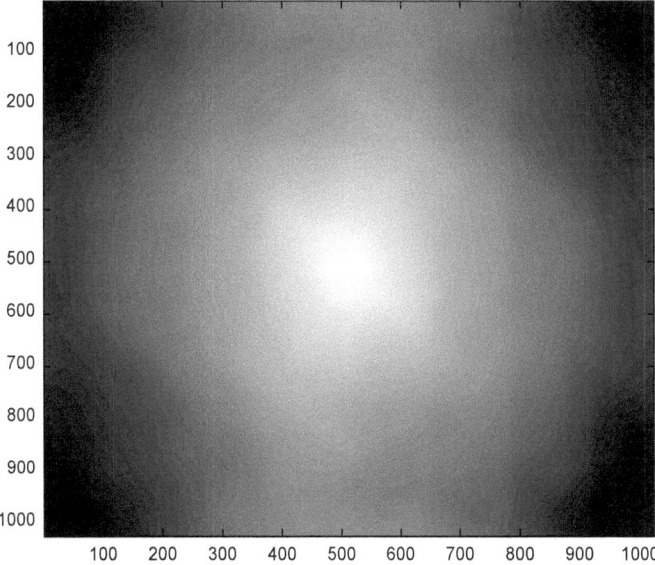

Fig. 11.8 Autocorrelation intensity of the heart image shown in Fig. 11.2

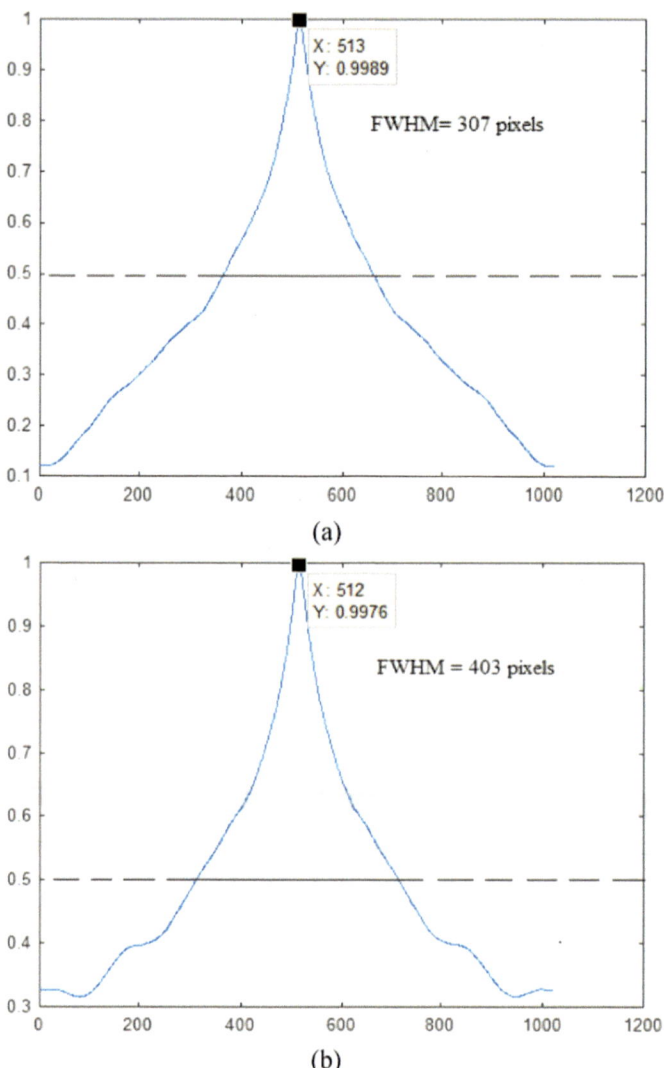

Fig. 11.9 a Line plot corresponding to the autocorrelation of heart at $x = 512$ pixels. **b** Line plot corresponding to the autocorrelation of heart at $y = 512$ pixels

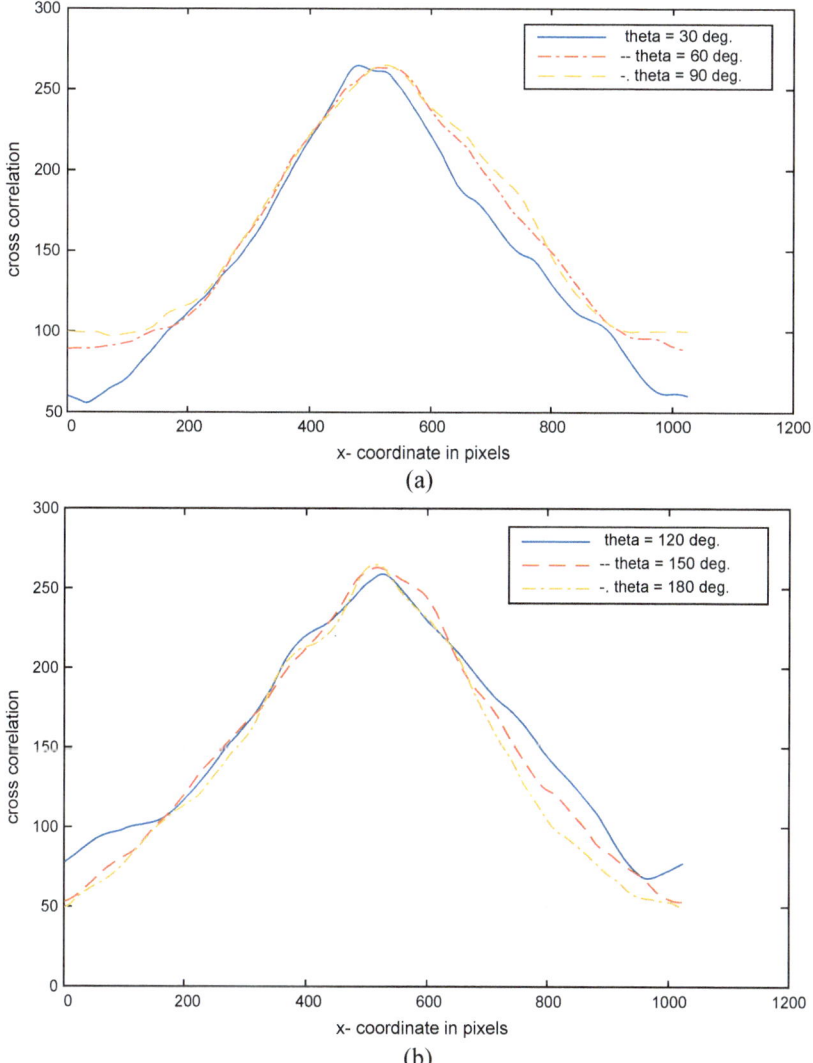

Fig. 11.10 **a** Cross-correlation corresponding to the unrotated heart image and the rotated images at angles of 30°, 60°, and 90°. **b** Cross-correlation at angles of 120°, 150°, and 180°

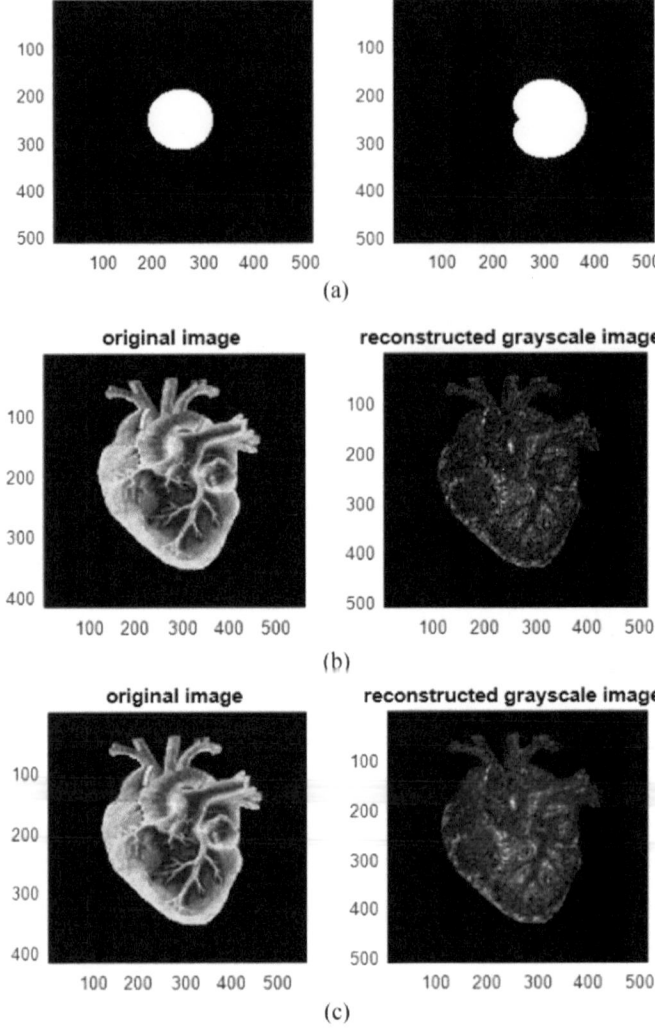

Fig. 11.11 **a** Apertures used for image acquisition via confocal microscopy. **b** The original cardiac image and its reconstructed image. The images have the same matrix dimensions of 512×512 pixels. The cardiac and circular apertures shown in (**a**) were used for reconstruction via confocal microscopy. **c** Original cardiac image and its reconstructed image. The images have the same matrix dimensions of 512×512 pixels. Circular apertures were used for reconstruction via confocal microscopy

References

1. C.G. Refael, R.E. Woods, *Digital Image Processing*, 2nd edn (Prentice Hall, Upper Saddle River, New Jersey 07458, 2001)
2. J.L. Semmlow, *Bio Signal and Biomedical Image Processing* (Marcel Dekker, Inc., 2004)

3. B. Jahne, *Digital image processing*, 5th edn. (Springer, Berlin, 2002)
4. A.V. Aho, J.E. Hopcroft, J.D. Ullman, *The Design and Analysis of Computer Algorithms* (Addison-Wesely, Reading, MA, 1974)
5. G.R. Arce, N.C. Gallagher, T.A. Nodes, *Median Filters: Theory for One-and Two-Dimensional Filters* (JAI Press, Greenwich, USA, 1986)
6. E.R. Dougherty, *Random Processes for Image and Signal Processing* (IEEE Press, NY, 2000)
7. E.K. Etienne, M. Nachtegael, *Fuzzy Techniques in Image Processing* (Springer, NY, 2000)
8. A.M. Hamed, Numerical speckle images formed by diffusers using modulated conical and linear apertures. J. Mod. Opt. **56**, 1174–1181 (2009). https://doi.org/10.1080/09500340902985379
9. A.M. Hamed, Formation of speckle images formed for diffusers illuminated by modulated apertures (circular obstruction). J. Mod. Opt. **56**(15), 1633–1642 (2009). https://doi.org/10.1080/09500340903277792
10. A.M. Hamed, Discrimination between speckle images using diffusers modulated by some deformed apertures: simulation. Opt. Eng. **50**, 1–7 (2011). https://doi.org/10.1117/1.3530085
11. A.M. Hamed, Image processing of Ramses II statue using speckle photography modulated by a new Hamming linear aperture. PRAM J. Phys. **94**, 126 (2020)
12. A.M. Hamed, The contrast of laser speckle images using some modulated apertures. PRAM J. Phys. **95**, 122 (2021)
13. A.M. Hamed, T. Al-Saeed, Image analysis of modified Hamming aperture: application on confocal microscopy and holography. J. Mod. Opt. **62**(10), 801 (2015). https://doi.org/10.1080/09500340.2015.1007102
14. A.M. Hamed, Improvement of point spread function (PSF) using linear quadratic aperture. Optik **131**, 838 (2017). https://doi.org/10.1016/j.ijleo.2016.11.201
15. A.M. Hamed, *Topics on Confocal Scanning Laser Microscope (CSLM)* (Lambert Academic Publisher, Germany, 2019). ISBN: 978-620-0-24595-3. www.lap.com
16. N.C. Pégard et al., Compressive light-field microscopy for 3D neural activity recording. Optica **3**, 517 (2016)
17. T. Seidel, J.C. Edelman, F.B. Sachse, Analyzing remodeling of cardiac tissue: a comprehensive approach based on confocal microscopy and 3D reconstruction. Ann. Biomed. Eng. **44**(5), 1436 (2016). https://doi.org/10.1007/s10439-015-1465-6
18. C. Huang, S. Wasmund et al., Catheterized fiber-optics confocal microscopy of the beating heart in situ. Cardiovasc. Imaging **10**, 10 (2017)